The Violent Technologies of Extraction

Alexander Dunlap • Jostein Jakobsen

The Violent Technologies of Extraction

Political ecology, critical agrarian studies and the capitalist worldeater

palgrave
macmillan

Alexander Dunlap
Centre for Development & the Environment
University of Oslo
Oslo, Norway

Jostein Jakobsen
Centre for Development & the Environment
University of Oslo
Oslo, Norway

ISBN 978-3-030-26851-0 ISBN 978-3-030-26852-7 (eBook)
https://doi.org/10.1007/978-3-030-26852-7

This Palgrave Pivot imprint is published by the registered company Springer Nature Switzerland AG.
The registered company address is: Gewerbestrasse 11, 6330 Cham, Switzerland

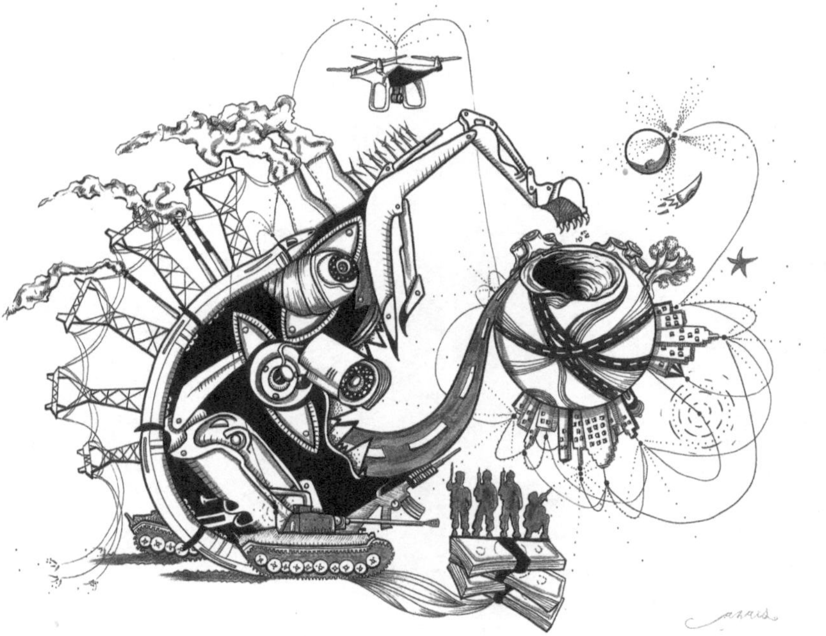

We dedicate this book to those—human and non-human—coming to daggers with the Worldeater, seeking to dismantle its rule for a social and ecologically harmonious future.

Acknowledgments

This book would not have been possible without the tireless efforts of the people close to us. Jostein would like to thank Ingfrid Knudegaard, Embla and Jo for allowing him to work on this project despite everything—thank you. We remain indebted to the thoughtful feedback from our three anonymous reviewers who offered us support, critical feedback and detailed scrutiny of our manuscript. This review work offered helpful guidance to complete this book—and the provocative ideas it contains. This also includes comments and suggestions on sections of this manuscript from Mariel Aguilar-Støen and Jaume Franquesa, who offered encouragement and the refinement of ideas. Finally, the Centre for Development and the Environment (SUM) at the University of Oslo with encouragement and moral support from Sidsel Roalkvam as well as Poul Wisborg, Kristoffer Ring, Benedict Bull, Kristian Bjørkdahl, Gudrun Cecilie Eikemo Helland, Anne-Line Sandåker and many others.

Contents

1 Introduction: Consuming Everything—Capitalism and the Imperative of Total Extractivism 1

2 The Spirit and Metaphysical Form of Capitalism: Devils, Worms, Octopuses and Worldeater(s) 13

3 Studying the Worldeater(s): Political Ecology and Critical Agrarian Studies and Their Origins, Differences and Convergence 43

4 Claws & Teeth: The Militarization of Nature 73

5 The Worldeater(s) in Process: Uncovering the Nexus of Conventional and 'Green' Extraction 91

6 Conclusion: Out of the Entrails—Reflections on Human Power 119

Bibliography 133

Index 161

List of Figures

Fig. 2.1 This is a depiction of Hobbes' Leviathan (Source: Wikipedia.org) 23

Fig. 2.2 Map of 'world energy regions' and 'world energy grid' (Source: The Energy Report by Metropolitan Office of Architecture (AMO)) 28

CHAPTER 1

Introduction: Consuming Everything—Capitalism and the Imperative of Total Extractivism

Certain human realities become clearer at the periphery of the capitalist system, making it easier for us to brush aside the commoditized apprehension of reality.
—Michael Taussig

The beast knows itself to be a machine, and it knows that machines break down, decompose, and may even destroy themselves. A frantic search for perpetual motion machines yield no assurances to counter the suspicions, and the beast has no choice but to project itself into realms of beings which are not machines.
—Fredy Perlman

Abstract The earth and its inhabitants are on a trajectory of cascading socio-ecological crisis driven by techno-capitalist development. Presenting the aim and scope of this book, the introduction lays out the key conceptual issue of *total extractivism*, naming the spirit and amalgamation of violent technologies comprising the totalizing imperative and tension at the heart of the present catastrophic trajectory. Total extractivism denotes how the techno-capitalist world system harbors a rapacious appetite for all life—total consumption of human and non-human resources—that destructively reconfigures the earth. Drawing on hostile, dissident authors and their companions—humans who have resisted techno-capitalism—the introduction sets the scene for viewing the Leviathanic capitalist state system and its expanding grid of extractive infrastructures as the Worldeater(s).

A. Dunlap, J. Jakobsen, *The Violent Technologies of Extraction*,
https://doi.org/10.1007/978-3-030-26852-7_1

Keywords Capitalism • Extractivism • Crisis • Extraction • Violence

Forests are replaced by plantations, estuaries with asphalt and water with synthetics liquids—chemical solvents or industrial wastes—that have technical names such as methylene chloride or arsenic. The rate of poisoning of the earth and its inhabitants is astounding, a rate corresponding to the progressive erosion and tokenizing of land-based practices and knowledges. Shocking, on the other hand, is the amount of scientific knowledge, measurement and debate of this destruction while it proceeds unabated and is normalized into ecosystems, daily life and the organisms of humans and non-humans. The question emerges: Why? How can the destruction of so much beauty and life continue? Drinkable water is turned into sewage and chemical run off; life is confined by concrete, steel and particle board; and interactions with nature are turned into a hobby to be sold as an identity. This book tries to make sense of this trajectory of 'progress'—as it is perversely called—and its continuation through the spirit of the Worldeater.

This book is a provocation. It is even a cry for help to consider the accumulative implications of the present socio-ecological trajectory and its ramifications on humans and non-humans in the age of anthropogenic climate change, species extinction and overall cascading ecological crises. To be more precise, according to the United Nations (UNSDG 2018), 'thirteen million hectares of forests are being lost every year'; meanwhile the 'degradation of dry lands has led to the desertification of 3.6 billion hectares.' Habit loss and soil degradation coalesces with the proliferation of roads, power lines, plantations, mines and factory farms that colonize rural areas and feed the rapidly growing urban populations that have increased 'from 751 million in 1950 to 4.2 billion in 2018' (UNSDG 2018). The world total energy supply has increased 60% since 1990 levels as mass consumption continues unabated (UNSDG 2019).[1] Energy production and consumption—industrial economic development—is at the heart of climate change, which is resulting in ocean acidification, 3-inch (7.62 centimeters) rises in sea levels since 1993, the increased frequency of extreme weather since the 1960s and the proliferation of large-scale forest fires since the 1980s (Wuebbles et al. 2017). Nature loss is rampant, leading the Intergovernmental Science-Policy Platform on Biodiversity and Ecosystem Services (IPBES 2019: 4) to assess that '[n]ature across most of the globe has now been significantly altered by multiple human drivers, with the majority of indicators of ecosystems and biodiversity showing rapid decline'. 'Progress', as it is popularly conceived, is radically discredited by these catastrophes. In short, Polanyi's (industrial) *Great Transformation* has come home to roost in the form of multilayered socio-ecological crises, but how does it continue? The kids are peacefully striking for climate change, the adults tweet about it and we—academics—sit at our computers and respective jobs while the world burns. Why? Is this the only way?

In his classic anthropological study examining 'peasants' or 'neophyte proletarians' employment of the figure of the devil to explain the onset of plantations and mines, Michael Taussig (1980) acknowledges how their perspective is belittled. The devil of extractive exchange-value is reduced to anxious emotional reactions—superstition—of 'savages' or written off as a social function or utility within rural communities. Unsatisfied with this blind discrediting of mythology that performs capitalist and/or extractivist apologetics, Taussig (1980: 17) asks: 'why not see them in their own right with all their vividness and detail as the response of people to what they see

[1] See https://yearbook.enerdata.net/total-energy/world-consumption-statistics.html.

as an evil and destructive way of ordering economic life?' Given the sustained trajectory of (classist, racist and patriarchal) industrial humanity and where it has placed itself, non-humans and the planet as a whole, we see great value in this approach. This book thus seeks to build on the social hostility of the people who have tried to make sense of their domination by governments, commodities and technologies—to list the broad signifiers of 'progress'—through mythical figures. In this tradition, we also seek to draw on "creatures" and "spirits" to understand a world on fire with peoples' complacency indicating their seeming eagerness to watch it burn—even celebrating the expanse of megaprojects—meanwhile researchers measure and fixate over abstract calculations of increasing ecocide, and ultimately appear puppeted under an industrial regime of asphalt, concrete, steel and enchanting electronics. Because remember, according to a recent (conservative) overview (Hickel and Kallis 2019: 7, 15), 'green growth theory—in terms of resource use—lacks empirical support' and at best 'remains a theoretical possibility' which means that there 'is no reason to design policy around it when the facts are pointing in the opposite direction'. Said simply, the techno-capitalist solutions are not working, or they only work to the extent that they are based on faulty scientific assumptions and reductive measurements, but this should not be surprising to a keen observer.

Distraught over the present state of the world—and even bitter in our own forced participation in it[2]—this book roots itself in the mythology and hostility of people appalled or in opposition to the techno-capitalist system. This is a theoretical perspective, discussed at length in Chap. 2, which views modernization and the euphemistic progress headed by the Leviathanic state—or state system—as geared toward systematically (1) affirming the organization of statist political economy (despite organizational restructuring and prioritizing); and (2) realizing the logic and imperative of *total* political and natural resource control. We have to question the purpose, reason and determination of universal categories—'state', 'society', 'corporation' to name a few—what they create and how they are assembled and re-organize themselves. While state organizations and various national and international institutional apparatuses exist, and comprise a complex and now computationally based (state) system, the

[2] While one might justifiably argue we have benefits from growing up in industrial societies, and continue to indulge in their pleasures, they were not of our choosing and have resulted in psycho-social and ecological costs, begging the question: do these costs collectively outweigh their benefits?

movement toward total resource control remains a tension and lived trajectory of industrial society. We see the state as the coercive-institutional framework that houses the economy and private sector often executing extractivist or mega-development projects, which together form the motor of techno-capitalist modernity that transcends political regimes. This book takes a distanced perspective—looking down from an airplane—at the formation and accumulation of larger socio-political and infrastructural outcomes. From this distance, the book extensively animates crucial disciplinary debates, specificities and changes in the larger process of techno-capitalist development, drawing extensively upon critical agrarian studies and political ecology, while bringing these into dialogue with dissident authors and thought (see Chaps. 2, 3, 4 and 5).

We call this imperative of the state system *Total Extractivism*. With this term, we build upon, while seeking to transcend what recent scholarship emanating from Latin America increasingly calls 'extractivism' or 'neo-extractivism' (Gudynas 2009; Lang and Mokrani 2013 [2011]). This emerging body of scholarship takes this concept to demarcate a twenty-first-century (re)turn to natural resource extraction as the motor of capitalist growth, with hyper-destructive ramifications that continues to position the Global South as the provider of raw materials in the international division of labor (Gudynas 2013 [2011]; Veltmeyer and Petras 2014; Aguilar-Støen 2016). Extractivism commonly 'refers to those activities which remove large quantities of natural resources that are not processed (or processed only to a limited degree), especially for export', explains Alberto Acosta (2013 [2011]: 62), going further to stress that extractivism is not confined to minerals and hydrocarbon but 'is also present in farming, forestry and even fishing' practices. Widening extractivism, Jaume Franquesa (2018: 145) locates three criteria that acknowledges 'green' extraction: 'centralization of ownership and decision making in a handful of major corporations, a productivist understanding of energy and concentration of production in a series of peripheral rural regions'. While debated primarily as a phenomenon in countries in the South, extractivism has always existed and is slowly escalating in the Global North (Brock and Dunlap 2018; Crosby and Monaghan 2018; Franquesa 2018),[3] which is taking on planetary relevance.

[3] The spread of fracking in the United States and England as well as Trump's opening of national parks to extraction serve as recent examples.

Recent work goes further to interrogate the broader relation between extractivism and global capitalism with its financialized logics and apparatuses (Gago and Mezzadra 2017). In one of the most stringent expositions to date, worth recounting at length, Ye and colleagues (2019: 2, 3) list ten defining features of extractivism, involving (1) 'the creation of a monopoly over the resources' that becomes (2) 'intertwinement between state and private capital groups'. This (3) requires infrastructural development, (4) often controlled by an 'operational centre' that (5) accumulates the generated wealth. Extractivism then (6) triggers inequalities that (7) sometimes entails various degrees of remediation by the state through redistribution. Extraction, most of all, (8) amounts to 'production without reproduction'—that is, destruction—amidst (9) 'boom-like' profits that (10) results in socio-ecological 'barrenness': degradated societies and ravaged landscapes. While such analytical specificity advances the debate, our discussion will make clear that we nevertheless disagree with the hyper-emphasis on centralization and see nuance in their 'production without reproduction'. Specifically, we see the 'rolling out' of decentralized logistical systems (see Cowen 2014; Hildyard 2016; Franquesa 2018), that while containing central nodes, remain diffuse and expansive in spreading industrial infrastructure, extractive sites (industrial and smallholder), communication and transport systems. Secondly, we see the disaster and degradation of extractivism reproducing new opportunities for 'green' economic initiatives ('new natures'), resorts and other methods of capitalist recuperation and intensification that profit from extractive disaster (see Dunlap and Fairhead 2014; Sullivan 2013; Brock and Dunlap 2018; Franquesa 2018), where the conventional and 'green' extraction nexus arises, which is discussed at length in Chap. 5. Moreover, in their eagerness toward defining extractivism vis-à-vis 'capitalism generally', Ye and colleagues (2019) could go further in probing the dark waters of techno-industrial 'progress' where extractivism finds its proper home.

Thus, we propose thinking of extractivism in a much broader way. Broader and, we would claim, deeper. *Total extractivism*, we argue in this book, is the imperative driving the global capitalist economy, centered on the deployment of violent technologies aiming at integrating and reconfiguring the earth and absorbing its inhabitants, meanwhile normalizing its logics, apparatuses and subjectivities, as it violently colonizes and pacifies various natures. While never complete—or fulfilled—and always resisted, appropriated and negotiated to various degrees, the techno-capitalist system indeed has the tension of *totalizing*. It harbors a rapacious appetite for all life, desiring the total consumption and reconfiguring of the earth centered on bureaucracy, industrial/cybernetic production and market relations that maintains a hyper-destructive growth imperative that produces a grotesque earthly product. Capitalism—past, present and future—attempts to devour all vitality: plants, animals, humans, hydrocarbons, minerals and just about anything 'seen', valued or revalued by the state and its appendages. Capitalism, in other words, colonizes the earth as it appropriates, expropriates and extinguishes the entirety of the earth's resources, which we discuss in greater depth in Chap. 2, drawing on Fredy Perlman's (2010 [1983]) notion of the *Worldeater* (see front pages), to name the amalgamation of violent technologies and spirit propelling the global capitalist economy toward total extractivism.

Through the prism of total extractivism and the Worldeater, this book guides the reader through the integration of otherwise seemingly disparate modalities of extraction that we define—in the broadest sense: to get, pull or draw out, usually with special effort, skill or force—within a systemic optic. These modalities and their violent technologies range from conventional extraction (mineral, oil, gas, etc.) to 'green' or 'renewable' extraction (agriculture, renewable energy and conservation). This book thus serves as a cursory intervention and map of *the organization of total extractivism or, said differently, the progress of the Worldeater.*[4] This progress, we acknowledge, proceeds at variegated intensities, sometimes more complete, other times with fewer layers and organs or centralized nodes, yet the progressive operation of worldeating continues. Chapter 2 begins with drawing up a theory of total extractivism, building extensively on subjugated or largely neglected knowledges from outside the academy. Here we draw on the iconoclastic Max Stirner (2017 [1845]) to understand the way 'phantasms'[5] or 'ghosts' were created through universal categories that gave rise to 'devils' (Taussig 1980), and formed other monstrous forces, such as Perlman's (2010 [1983]) giant mechanical worms and octopuses that fought, morphed and reproduced to create Leviathanic Beasts: The Worldeater(s). While there is a unitary drive to create a Worldeater, we acknowledge the possibility of various Worldeater(s)

[4] While this short book offers a wide-ranging mapping exercise, we acknowledge its limitations—and the limitations of our knowledge—suggesting instead that readers interested in exploring total extractivism further may want to interrogate works on aquatic resources or narcotics as key frontiers to extractivist expansion, or think of digital media technology as a key violent technology at its service—to name but a few possible avenues.

[5] In earlier translations, called 'spooks'.

operating independently, if aiming for the same goal of material formation. The central question regarding Worldeater(s) origins in human (pre-) history remains open: whether autonomous spirit(s)—so-called Devils—already existed and/or emerged from another realm to possess humans or alternatively—as we contend building on Stirner—are the products of humans—specifically their collective fears and insecurities manifesting into material and infrastructural form. For this purpose, the book develops these beasts and their logic through various authors—Jacques Ellul (1964 [1954]), Peter Gelderloos (2017) and Zapatistas (2016) to name a few—to understand the beastly spirit of capitalism (to give a twist to Weber's famous thesis) behind total extractivism.

Dissatisfied with terms such as 'climate change', 'crisis' or even the Anthropocene to describe the techno-capitalist transformation engulfing the planet, we find climate change as a technocratic framing that situates agency in the hands of climate scientists, governments, corporations and NGOs as inadequate to confront the evolving catastrophe at hand (see Dunlap 2016; Mann and Wainwright 2018). Crisis, meanwhile, has become a permanent and normalized feature of techno-industrial society that, we can say, is as natural as capitalism itself. In a real sense, we are governed by 'crisis' (see TIC 2015), demanding us to scrutinize alternative framings. While we remain critically supportive of these prevailing terms (and their theoretical development), we respond to this need for alternative framings as we draw upon Perlman and others to develop a dissident (anarchistic) approach that departs from much existing academic work.

The book then takes a more disciplinary—one is tempted to say, disciplined—academic turn in Chap. 3. We recognize political ecology and critical agrarian studies as central to developing key insights into extractivism and socio-ecological catastrophe, laying more or less implicit foundations for the Worldeater approach presented in this book. However, we would also like to offer the concept of the Worldeater as an invitation for political ecology and critical agrarian studies to actually employ, criticize and develop further. This means accentuating the radical anarchistic and pluriversal tendencies already existing within these fields to resurge discussion and praxis in a world confronted by the Worldeater's imperative toward total extractivism. This resurgence may, we suggest, enable the disciplines to expand their view, also, of liberation—toward total liberation. Turning to these key academic studies or subfields, Chap. 3 seeks to bring together the varied and, too often, disconnected subfields to begin mapping and charting the movements, functions and phenomena that, we would say, are the 'springs and wheels' of the Worldeater and its expansive strategies. We analyze political ecology and critical agrarian studies, specifically with their concerns with political power, land and resource relationships and control to chart their commonalities, differences but also weaknesses in analyzing the onset of techno-capitalist progress or the Worldeater. Spanning across what is popularly and all-inclusively called 'environmental humanities', critical agrarian studies and political ecology—as they relate to anthropology, geography and development studies—retain the greatest possibility for analyzing the parts, processes and whole of the Worldeater.

Central to the violent technologies of extraction is coercion and social pacification. Chapter 4 discusses the centrality of militarization and violence—'the claws and teeth' of the Worldeater(s)—in rolling out and enabling extractivism and the resulting environmental degradation. This is a discussion of violence and social control that is necessary to perform the tasks of extractivism, and to discipline and convert the subjectivities of people to relinquish their relationships with their lands and ecosystems. This violence, however, is myriad and takes many forms, not always involving armored vehicles, riot police and helicopters, but the articulation of strategies of social pacification that involve, following Foucault (2007 [1978]), 'positive mechanisms' that enchant and allure populations to engage in the spoils of capitalism. This violence and evolution of techniques, however, is complex and takes on increasingly devious forms when so-called security forces operate under the guise of environmental protection and regulation. This overview seeks to animate the violent technologies of the Worldeater and the regimenting of total extraction into people and landscapes, presenting detailed insights into the violence of land control and territorialization and the political ecology of counterinsurgency.

While Chap. 4 examines conventional and 'green' militarization, Chap. 5 examines the relationship between conventional and green extraction industries. This section examines the shifting and blurring lines between conventional extraction associated with mineral and hydrocarbon extraction with renewable systems: agriculture, plantations, wind and solar industries.[6] We identify a crucial nexus between these modalities of extractivism—a nexus that is strongly implicated in the imperative of total extractivism and animates the progress of the Worldeater. The purpose here is to examine the subtle shifts and movements that animate and com-

[6]While plantations do not appear particularly 'green' in their exhaustion of the soil, this exhaustion is not complete in the same sense as with conventional extraction.

prise the push toward total extractivism, demonstrating how, contrary to popular belief, green or renewable energy industries are in fact spreading similar, if not the same, forms of ecological degradation—or more complicated forms thereof—and vital consumption. Chapter 5 also, and simultaneously, explores the ways that smallholders come to be inserted into the expanding grid of mines and plantations, enabling the integration of new rural areas into the reach of the rapaciously devouring machinery. In outlining this systemic colonizing grid, we emphasize the ways it is devised *with* and *through*—as opposed to *by*—humans. The dead thing—infrastructure—feeds on, and lives through, us. It is 'vampire-like', as Marx (1982 [1867]: 342) described capital, and it 'only lives by sucking living labour, and lives the more, the more labour it sucks.' Through the prism of total extractivism, this book offers a sustained provocation into the forces that continue to consume, reconfigure and spread capitalism, its infrastructure and mentality across the earth. This desperately begs the question: How can humans avoid possession by the Worldeater—or, if it is already too late, how can they escape from, as Perlman (2010 [1983]) calls it, its 'entrails'? To be precise—which is more pressing now than ever before: *Do humans have the agency to change the trajectory of total extractivism?*

Reflecting on previous chapters, this is the discussion taken up in Chap. 6, the Conclusion, where we outline what we believe needs to be resisted and, in broad strokes, how. Offering a general projection and ethos to undermine the smooth functioning of the Worldeater, we do not believe in the myths of a 'technological fix' all-too-prevalent in the justifications of eco-modernism to renew the present with its fraudulent shades of 'green'. Total extractivism, we argue in the Conclusion, demands nothing less than the accelerated uptake of *total liberation* as an ethos and political stance that integrates the totality of hierarchical relationships across human beings and other species that remains grounded in people and their socio-historical context, yet agrees on a general trajectory of how to live in this world. While we do not perceive total liberation as a fixed 'solution', we offer the notion more as a set of tentative guidance to begin to erode and decompose the Worldeater—we must compost the techno-capitalist colonial system. This approach, however, depends entirely on the determination of people, their commitment to the animals, trees, rivers and friends where they live or travel—in a word: habitat. This is a generational commitment, a struggle older than we are and a struggle where we only have ourselves and each other to subvert the Worldeater and its growth imperative toward total extractivism.

Bibliography

Acosta, A. (2013 [2011]). Extractivism and Neoextractivism: Two Sides of the Same Curse. In M. Lang & D. Mokrani (Eds.), *Beyond Development: Alternative Visions from Latin America* (pp. 61–86). Amsterdam: Transnational Institute.

Aguilar-Støen, M. (2016). Beyond Transnational Corporations, Food and Biofuels: The Role of Extractivism and Agribusiness in Land Grabbing in Central America. *Forum for Development Studies, 43*, 155–175.

Brock, A., & Dunlap, A. (2018). Normalising Corporate Counterinsurgency: Engineering Consent, Managing Resistance and Greening Destruction Around the Hambach Coal Mine and Beyond. *Political Geography, 62*, 33–47.

Cowen, D. (2014). *The Deadly Life of Logistics*. Minneapolis, MN: University of Minnesota Press.

Crosby, A. C., & Monaghan, J. (2018). *Policing Indigenous Movements: Dissent and the Security State*. Black Point: Fernwood Publishing.

Dunlap, A. (2016). Counter-Insurgency: Let's Remember Where Prevention Comes from and Its Implications. *Critical Studies on Terrorism, 9*, 380–384.

Dunlap, A., & Fairhead, J. (2014). The Militarisation and Marketisation of Nature: An Alternative Lens to 'Climate-Conflict'. *Geopolitics, 19*, 937–961.

Ellul, J. (1964 [1954]). *The Technological Society*. New York: Vintage Books.

Foucault, M. (2007 [1978]). *Security, Territory, Population: Lectures at the College de France 1977–1978*. New York: Picador.

Franquesa, J. (2018). *Power Struggles: Dignity, Value, and the Renewable Energy Frontier in Spain*. Bloomington, IN: Indiana University Press.

Gago, V., & Mezzadra, S. (2017). A Critique of the Extractive Operations of Capital: Toward an Expanded Concept of Extractivism. *Rethinking Marxism, 29*, 574–591.

Gelderloos, P. (2017). *Worshiping Power: An Anarchist View of Early State Formation*. Oakland, CA: AK Press.

Gudynas, E. (2009). Diez tesis urgentes sobre el nuevo extractivismo. In J. Schuldt, A. Acosta, A. Barandiarán, et al. (Eds.), *Extractivismo, política y sociedad*. Quito: Centro Andino de Acción Popular (CAAP) and Centro Latinoamericano de Ecología Social (CLAES).

Gudynas, E. (2013 [2011]). Debates on Development and Its Alternatives in Latin America: A Brief Heterodox Guide. In M. Lang & D. Mokrani (Eds.), *Beyond Development: Alternative Visions from Latin America* (pp. 15–40). Amsterdam: Transnational Institute.

Hickel, J. L., & Kallis, G. (2019). Is Green Growth Possible? *New Political Economy*. https://doi.org/10.1080/13563467.2019.1598964.

Hildyard, N. (2016). *Licensed Larceny: Infrastructure, Financial Extraction and the Global South*. Manchester: Manchester University Press.

IPBES. (2019). Report of the Plenary of the Intergovernmental Science-Policy Platform on Biodiversity and Ecosystem Services on the Work of Its Seventh Session [Advance Version]. Retrieved from https://www.ipbes.net/system/tdf/ipbes_7_10_add-1-_advance.pdf?file=1&type=node&id=35329.

Lang, M., & Mokrani, D. (2013 [2011]). *Beyond Development: Alternative Visions from Latin America*. Amsterdam: Transnational Institute.

Mann, G., & Wainwright, J. (2018). *Climate Leviathan*. London: Verso Books.

Marx, K. (1982 [1867]). *Capital: A Critique of Political Economy* (Vol. I). London: Penguin Books.

Perlman, F. (2010 [1983]). *Against His-Story, Against Leviathan*. Detroit: Red and Black Press.

Stirner, M. (2017 [1845]). *The Unique and Its Property*. Berkeley, CA: Ardent Press.

Sullivan, S. (2013). After the Green Rush? Biodiversity Offsets, Uranium Power and the 'Calculus of Casualties' in Greening Growth. *Human Geography, 6*, 80–101.

Taussig, M. T. (1980). *The Devil and Commodity Fetishism in South America*. Durham, NC: University of North Carolina Press.

TIC (The Invisible Committee). (2015). *To Our Friends*. South Pasadena: Semiotext(e).

UNSDG. (2018). *Sustainably Manage Forests, Combat Desertification, Halt and Reverse Land Degradation, Halt Biodiversity Loss*. Retrieved from https://www.un.org/sustainabledevelopment/biodiversity/.

UNSDG. (2019). UN Report: Nature's Dangerous Decline 'Unprecedented'; Species Extinction Rates 'Accelerating'. Retrieved from https://www.un.org/sustainabledevelopment/blog/2019/05/nature-decline-unprecedented-report/.

Veltmeyer, H., & Petras, J. F. (2014). *The New Extractivism: A Post-Neoliberal Development Model or Imperialism of the Twenty-First Century?* London: Zed Books.

Ye, J., van der Ploeg, J. D., Schneider, S., et al. (2019). The Incursions of Extractivism: Moving from Dispersed Places to Global Capitalism. *The Journal of Peasant Studies*, 1–29.

Wuebbles, D., Fahey, D., Hibbard, K., et al. (2017). *US Global Change Research Program: Climate Science Special Report*. Retrieved from https://science2017.globalchange.gov/chapter/executive-summary/.

Zapatistas. (2016). *Critical Thought in the Face of the Capitalist Hydra I*. Durham: Paper Boat Press.

CHAPTER 2

The Spirit and Metaphysical Form of Capitalism: Devils, Worms, Octopuses and Worldeater(s)

Societies that dominate nature also dominate people. Where there is the idea that a massive dam should be built to control a river's flow, there is the idea that people should be enslaved to build it; where there is the belief that a giant metropole may serve itself by despoiling the surrounding countryside and devouring its raw materials, there are castes and hierarchies to ensure that this is accomplished.
—Kirkpatrick Sale, 1985

Abstract This chapter seeks to answer how and why the world is being consumed, digested and ultimately reconfigured. It presents the concept of the Worldeater(s) in its manifold forms through an extensive theoretical discussion of the metaphysics—spirit—and shape-shifting processes underpinning techno-capitalist development. While unconventional in academic circles, the dissident perspectives drawn upon reveal some of the ways that the trajectory of total extractivism normalizes and continues unabated as human beings and all other life forms are converted into entrails of the Leviathanic beast. Drawing extensively on writers such as Max Stirner, Michael Taussig and Fredy Perlman, this chapter uncovers the roots of the Worldeater(s) and its colonizing trajectory across the earth. The chapter shows how this colonization is made possible and furthered by *techniques*—the logic of the violent technologies of extraction.

A. Dunlap, J. Jakobsen, *The Violent Technologies of Extraction*,
https://doi.org/10.1007/978-3-030-26852-7_2

Keywords Capitalism • Extraction • Anarchism • Colonization • Violence • Technology

The world is being transformed, converted and consumed. The etymological roots of the word 'colonialism' are conventionally traced to the Latin *colonia* meaning 'settled land' and *colonus* meaning 'settler in new land' (Dunlap 2018b). However, an intriguing—and highly subversive—alternative rooting is found in the word *colon*: the large intestine. This would mean that 'colonialism is the digestion of one people by another—'the body politic ("the state") works to digest and "assimilate" the Indigenous peoples [and others] that it "eats," and "absorbs"' (Platt 2013; Dunlap 2014: 61). One could argue that colonization is natural or—the indirect version—that some type of extraction is always necessary: humans are always eating, killing and consuming some aspect of non-human nature. This type of sentiment, however, articulates a type of colonial apologetics, neglecting not only the relationship, but more importantly for our purposes here, the scale of consumption and, more so, the product:

the excretion of that digestion—the stool, the byproduct or the 'shit' colonial systems produce. As we will elucidate below, we see great structural affinity between the civilized, colonial and statist model and what it produces.[1] Urine, stool, blood and, not to forget, corpses—even if they can be great sources of disease if killed in mass and/or not (properly) buried—themselves are essential ecological resources, offering vital nutrients for flora and food for fauna that animates and replenishes the so-called web of life. Now, what is the new cycle or, in mechanical philosophical terms, 'system' being produced today? What has colonialism and statist forms of organization produced with their consumption of peoples—do their bodies return to the animals and soil?[2] In one sense, what is produced are amazing automotive and flying machines, computers, vaccines, processed foods and, in a word, technological wonder that was once only a product of science fiction. In another sense, this 'wonder' has produced widespread diseases, illness, industrial disasters, uranium babies and, in short, social, ecological and climate catastrophe. In all fairness, these negative outcomes of modernization and capitalist growth are turning out to be extremely profitable for some industries and people, but that is assuming technological enchantment and profit is your bottom line. As Paul Virilio (2008 [1983]) summarized it: 'The invention of the boat was the invention of shipwrecks'. We might say, hindsight is 20/20, yet this trajectory has always been imposed by various means and resisted from the inside as well as the outside from ancient civilizations until present.[3]

While there might be appropriate forms of landscape colonization, its byproducts should not be genocide, ecocide and ever evolving forms of slavery. As every indigenous culture is well aware, humans are nature; humans are intimately implicated in and part of the environment—not separate. Murray Bookchin (1982), Élisée Reclus (2013 [1905]) and Kirkpatrick Sale (1991 [1985]), quoted above, as well as ecofeminists, deep ecologists and many more (Thoreau 1971 [1854]; Naess 1973;

[1] See Dunlap (2018b) for definitions of colonialism and the colony in relationship to the state.

[2] Yes, but often in concrete caskets and filled full of preservatives.

[3] If this statement seems odd, please consider reading Linebaugh P and Rediker M. (2013) *The many-headed hydra: sailors, slaves, commoners, and the hidden history of the revolutionary Atlantic*, New York: Beacon Press; Merchant C. (1983) *The Death of Nature: Women, Ecology, and The Scientific Revolution*, New York: Harper & Row; Sakolsky RB and Koehnline J. (1993) *Gone to Croatan: Origins of North American Drop Out Culture*, New York: Autonomedia.

Fukuoka 2010 [1978]; Merchant 1980; Mies and Shiva 2014 [1993]; Springer et al. 2019) were all too aware of these circular and self-reinforcing ripples of control and domination that disrupt the web of life. If imbalance and disruption are allowed to persist long enough, as we will see—and are seeing—it threatens all the life of the planet, and potentially the planet itself—the planet might end up like Mars, a landscape largely inhospitable to humans. Yet, myths of human supremacy, fixation with mechanical philosophy and computational systems continued this trajectory of progress (see Duffield 2011; Dunlap 2014; RF 2016; Jensen 2016). What is this system? Where did it come from and why, in all of its enchantments, is it continuing to consume the world? Is this the result of agriculture, bureaucratic planning, divisions of labor, wealth or technology? Likely all play their part, yet important here, and following Taussig (1980), is regaining the medium by which we communicate this process, outside the terms and logic of market relationships.

The politico-economic situation has normalized. 'In fact it appears so natural that the issue of dominance rarely arises', explains Taussig (1980: 28), 'in this sense the commodity form has truly subjugated the consciousness of persons who are endowed with a long capitalist heritage, but not, it would seem, the consciousness of those peasants with whom we are concerned—persons just beginning to experience capitalism'. The Indigenous or peasant cultures that encounter capitalist culture, Taussig (1980: 28) observed, would begin 'anthropomorphizing their subjugation'—in this case the figure of the devil is responsible for plantations, mines and the labor conditions to which they are subjugated and/or submit. Yet, this anthropomorphizing and means of diagnosing 'capitalist culture', Taussig (1980: 29) contends:

> [I]s living testimony to the legacy of the ideology through the ages that has assailed market exchange as something unnatural—a social form that undermines the basis of social unity by allowing creativity and the satisfaction of need to be subverted by a system that puts profit seeking ahead of people and that makes man [sic] an appendage of the economy and a slave to the work process instead of the master of it.

It is entirely in this tradition that we wish to proceed, framing this inquiry in mythic expression to convey the detrimental situation the earth is facing. This is a challenge to technocratic and academic language that languishes in meandering socio-ecological disaster, meanwhile embodying and projecting a habitus antithetical to socio-ecological change, whether from excessive plane flights to lifestyle organized by computers, but more so collectively acquiescing to banal and energy intensive infrastructural environments that embody the perpetuation of ecological catastrophe every day. This is not to take a righteous position, but to recognize the depths and normalization of the issues at hand. 'Crisis', 'climate change' and 'Anthropocene', while important terms, we believe are inadequate to describe what is happening to the earth and its inhabitants—indeed, these terms may, when inserted in dominant technocratic vocabularies and apparatuses, perpetuate this crisis.[4] Charting a novel approach, we respond by drawing on diverse authors to offer a mythological conception of the spirit and metaphysical form of capitalism.

Drawing on dissident authors, below we seek to answer how and why the world is being consumed, digested and ultimately reconfigured, before returning to academic review prose of documenting the intricacies that make up the process of *worldeating* (Chaps. 3, 4 and 5). The conclusion will then take up the question of human agency in the face of industrial human-made socio-ecological catastrophe. Let this be a playful—while serious—exercise in the insurrection of subjugated knowledges (see Foucault 2003 [1997]), and in terms largely frowned upon in academia—story and myth—that better convey a reality we have grown accustomed to and desire. Despite this enchantment, paradoxically, people still feel the negative effects in their bones.[5] That is, if we can still feel that pain in spite of the self-medication and prescription of drugs, alcohol and blinders we create to normalize and adjust to industrial life. Below is another story about what is happening to the world and its inhabitants.

[4] You decide. This is a self-reflective cry against the present organization and regime of knowledge production.

[5] Anti-depressants' consumption has increased 65% in the last 15 years in the United States (Pratt et al. 2017), meanwhile doubling in OECD countries between 2000 and 2015 (OECD 2017: 189–91). This resonates with depression, psychological distress and suicides that increased 30% overall and 50% among women in the United States between 2000 and 2016 (Hedegaard et al. 2018).

From Ghosts to Worldeaters: The Mythic Reality Hidden in Plain Sight

We propose a radical starting point to the origin story of the Worldeater, which begins with taking Max Stirner (2017 [1845]) more seriously than is commonly done.[6] An iconoclastic and insurrectionary author, and ardent adversary of Marx (see Derrida 2012 [1993]) and critic of Proudhon, Stirner (2017 [1845]) has a cunning way of getting to the roots of domination and enslavement. In dissecting the construction of individual subjugation, Stirner (2017 [1845]) narrates how ghosts and phantasms are discursively born. Ghosts and phantasms, in Stirner's prose, are similar to Polanyi's (2001 [1944]) 'fictitious commodities'—land, labor and money—and can thus be subject to Taussig's (1980: 30) rejoinder to Polanyi: 'Fictitious indeed! But then how does one explain the persistence and strength of this fiction?' According to Stirner (2017 [1845]), there are even greater fictions that give rise to ghosts. These ghosts arise from the fiction and lived embodiment of universal categories that people believe, create and project by various psycho-social means and material acts. Such ghosts are the idea of 'society', 'tribe', 'family', 'the people' and the notion of 'humanity' itself. Dripping in disdain and sarcasm, Stirner (2017 [1845]: 197–8) elaborates:

> [H]uman beings have not yet been able to base their societies on *themselves*; or rather, they have only been able to found "societies" and to live in societies. These societies were always persons, powerful persons, so-called moral persons, i.e. ghosts, before which the individual had the appropriate bat in his belfry, the fear of ghosts. As such ghosts, they can most properly be called by the respective names "people" and "tribe;" the people of the patriarchs, the people of the Hellenes, etc., finally, the human people, humanity, then every subdivision of this "people," which could and must have its particular societies, the Spanish, the French people, etc.; within these as well, the estates, the cities, in short all kinds of corporations, last, at the extreme point the small tribe of the—*family*. Instead of saying that the person haunting all societies up to now has been the people, hence also the two extremes could be named, namely either "humanity;" or the "family," the two "most

[6] Speculating on even more radical starting points, we find ourselves venturing into the uncharted terrain of shamanistic warfare that, while enigmatically relevant, presents itself beyond the limits of our knowledge. Readers brave enough to trespass into such terrain may want to seek guidance from Clastres et al. (1977) Taussig and your local Kung Fu teachers versed in the methodology of their art and origins of tyrannical dynasties.

> natural units." We choose the word "people" because its origin has been brought together with the Greek *Polloi*, the "many" or the "mass," but more so because 'national aspirations' are at present the order of the day, and also because even the latest rebel has not yet shaken off this deceptive person, although, on the other hand, the latter consideration would have to give the advantage to the term "humanity," because on all sides people are starting to rave over 'humanity'. (emphasis original)

The broad signifier makes ghosts, it subjugates and makes political levers by which to control and organize populations according to 'national aspirations'. These ghost haunt, reconfigure, separate and, most importantly, bind people to the 'nation' and the new organization of the state—the Leviathan. These disembodied spirits—or fictions—that civilized humanity lets run through them remain 'tortured of a thought that can't create a body for itself', risking that it 'melts away into nothing' (Stirner 2017 [1845]: 203). The fictions, the ghosts—they want bodies; they want a physical form on the plain of humans and will construct themselves by the means at their disposal.

The critique of language, signifiers and (profound) reification has long since been developed (Zerzan 1988, 2005 [1999]; Sayer 1987), yet let us imagine that what is guiding humanity into social, ecological and climate crises are spirits, created and affirmed by humans. Taussig (1980) tells us that the devil did not emerge until the Middle Ages and not until the arrival of plantations and mines in southwest Columbia, but could this Devil be the result of various fictions? We can imagine the Devil, with all the duality this can imply, as a monolithic creature or more accurately Devil(s): an amalgamation of spirits, tempted by the same desire—maybe competing and playing with each other—to 'create a body' and manifest in the material world. Maybe to hold dominion over the inhabitants of the earth, maybe to exist in this world or possibly propelled by an envy of the abilities that mortals have? We might consider boredom as a motive in a purposeless existence where gambling, playing the game of domination or self-made entertainment is the pastime of choice? We do not know—the question remains open—yet keeping Stirner's phantasms in mind, we turn to explore the formation of what Fredy Perlman called the Worldeater.

Worms, Octopuses and the Beast: Against Leviathan Revisited

This World System—this techno-capitalist industrial system—is in fact a monster, and paying tribute to Max Weber (1992 [1930]), is a spirit manifesting itself through mines, plantations, concrete, steel and fiber optic cables. The spirit instantiates power, yet '[n]o one sees it because it is in plain sight, all the time—in the form of a high-voltage line, a freeway, a traffic circle, a supermarket, or a computer program', write The Invisible Committee (2015: 84). If this power is hidden, however, 'it's hidden like a sewage system, an undersea cable, a fiber optic line running the length of a railway, or a data center in the middle of a forest. Power is the very organization of this world, this engineered, configured, *purposed* world.' We cannot agree more with this literal reading of power and, despite the real benefits and allure from these infrastructural systems, we must recognize their extreme break with environmental harmony, incessant landscape degradation and constant need for energy in the form of human power/labor and electric generation (among secondary consumption processes such as food production) that form a structure of conquest—not mutual self-affirmation of human and non-human life cycles. This structure of systemic degradation, as we discuss below, articulates a prison society, but also a spirit of power bent on breaking socio-ecological harmony for the creation of its own body or, said differently, techno-capitalist progress.

Discussing Giovanni Arrighi's (1994) *The Long 20th Century* that offers a detailed treatment of the historical evolution of capitalism, Alex Gorrion (2013: 3) presents an appreciative critique:

> The proposition that capitalism is antithetical to the market sounds suspiciously reminiscent of Proudhon. And Arrighi's dialectical model of capitalist powers that tend towards alternating territorialist and then capitalist strategies of accumulation bears a lot in common with Fredy Perlman's model of Leviathan that constitutes itself now as a worm, now as an octopus. In simpler terms and admittedly less sophistication, and without supporting statistics, Perlman provides (eleven years earlier) a similar analysis. *Against Leviathan*, however, is much more sweeping than *The Long 20th Century*, as Perlman recounts the development of civilization going back thousands of years, and despite some factual flaws comes much closer to *capturing the spirit of power and accurately describing how it functions*, a task at which Arrighi with all his statistics falls woefully short. (emphasis added)

Confronting the Marxist erasure of anarchist theories and/or interventions (see also Springer 2016a, 2017), Gorrion could not have described better this juxtaposition of works. First, by recognizing that the issue of capitalism—and, consequently, we may add: present socio-ecological devastation—is far more temporally (as well as socially and psycho-emotionally) profound with roots in Ancient Civilization; and, secondly, acknowledging Fredy Perlman's (2010 [1983]: 189) pioneering work with its flaws, but more so its poetic intelligence in their description of 'the spirit of power' that possesses this world or, in his words, is eating it. Perlman calls this spirit of power a 'Worldeater'.

The implicit contention is that the spirit dominating the planet and its inhabitants is much older than capitalism or colonialism, but civilization itself. Perlman (2010 [1983]) is accredited as a forefather of anti-civilization thought (Zerzan 1988), locating the oppression of the present in ancient civilization(s). From Chinese dynasties, Mediterranean to Mesoamerican civilizations, they bare the seeds and the logic of the present colonial/state system organized around various forms: human and non-human hierarchies, speciesism (human supremacy), symbolic culture, patriarchy and divisions of labor/specialization that form the basis of industrial society, but also its alienation, inequality and ecological degradation (see GA 2005). Often positioning itself in antagonism to university systems, anti-civilization thought has various Marxist, primitivist and eco-anarchist articulations,[7] where Perlman (2010 [1983]) remains a central inspiration.

These spirits and devils emblematic of civilized development were geographically dispersed, taking semi-autonomous and multifarious developments of their own. Adapting geographically, culturally, psychologically and with every particularity in between, the will to dominate, control and merge as an omnipresent power arises. Building on Perlman (2010 [1983]), Peter Gelderloos (2017) details the techniques of state formation and governance employed to root the colonial project: The manufacturing of leaders; various divide and conquer strategies; symbolism and spatial domination tactics to name only a few. While Gelderloos explains the mechanics, Perlman tells us about the spirits, their consumption patterns and material development. This phantasm, according to Perlman (2010 [1983]: 27), first takes shape as 'a giant worm, not a living worm but a carcass of a worm, a monstrous cadaver, its body consisting of numerous segments, its skin pimpled with spears and wheels and other technological implements'. This grotesque and mechanical worm is the organization of civilized infrastructure: temples, monuments, buildings, walls and enclosures, as they spread across the countryside by war and

[7] See Green Anarchy Magazine (2002–2009), Return Fire (2013–Present), Black Seed (2014–Present), Landstreicher (2009), and Fitzpatrick (2018).

subterfuge. What animates and drives this worm is human and non-human life, most visible in the organization of warfare, but always dependent on domestication of land and vitality to propel the worm's formation and appetite. Absorption and conversation into civilized life, Perlman (2010 [1983]: 31) explains, is subordinating life 'into the entrails of an artificial worm's carcass'. Civilizing and proletarianizing, we could say more expansively, is the process of '*entrails making*'. In this 'process, individuals degrade into the genital organs of Capital', as Byung-Chul Han (2017: 5) describes. Similar to proletarianization, entrails making acknowledges the totality of this progressive (and incomplete) conversion through gender, class, 'race', bureaucratic and technological enchantment into the operations of the Worldeater.

The worm, however, is not alone. There is also the octopus. The worm reproduces: 'segments of the decomposed worm remain scattered over the countryside, and each segment tends to recompose itself into a complete worm', Perlman (2010 [1983]: 43) observes, 'Dead things have powers living beings lack'. Worms fight each other, mutate and develop. Furthermore, every new worm 'has accessories its predecessors lacked' (Perlman 2010 [1983]: 46). Arising from particular 'segments' of 'decomposed' worms was a monster 'reconditioned into mobile, octopus-like monstrosities that will transport' and reach 'places far beyond the stationary worm' (Perlman 2010 [1983]: 46). With the onset of 'developed' civilizations, the world is experiencing the spawning of different types of worms and octopuses, more adept at overseas commerce and colonization. The worms and octopuses battle for territory, divide and conquer each other as well as collaborate—not all that different from corporations (see Dugger 1989)—slowly merging into one. This push toward collaboration and centralization, Perlman (2010 [1983]: 193) calls the formation of a 'beast' that eventually becomes the Leviathan—the state. '[W]ormlike and octopus like Leviathans could be distinguished from one another, although the distinction began to blur already in the Islamic world', explains Perlman (2010 [1983]: 193). 'In the west, the two forms of Leviathan become so intertwined that it becomes impossible to characterize the Western Leviathan as either one beast or the other. The beast of the West is something the world has never seen before'. This is Leviathan: Hobbes applauds and celebrates the worms and octopuses of the world as increasingly demarcated, bureaucratic and technically advanced states begin to emerge and mediate the 'war of all against all', that as we know now, was just a projection of the world created by worms and octopuses that Hobbes embraced through his fear and/or insecurity generated by the world they created.

Fig. 2.1 This is a depiction of Hobbes' Leviathan (Source: Wikipedia.org)

Hobbes' doctrine is explicit about the collection of bodies and souls to animate Leviathan, the state (see Fig. 2.1). Crucial, however, is *how* people are absorbed. Perlman (2010 [1983]) relies heavily on conquest and coercion as a means of absorption, meanwhile acknowledging the widespread human refusal of—and flight from—the Leviathan's entrails. Though couched in a different language, James Scott's (2017) recent work on early state formation can be seen as broadly compatible with Perlman's more mythic narrative, as Scott comes close to spelling out the movements of the worm as it expands through warfare, slavery, control and subjugation of recalcitrant humans and non-human natures. Early states, Scott (2017: 116–7) writes, were based on the 'control and appropriation' of humans, non-humans and grains that were accordingly 'parasitized' through continuous expansions across territories.

> To see the early states as "population machines" is not far off the mark, so long as we appreciate that the "machine" was in bad repair and often broke

> down, and not only because of failures in statecraft. The state remained as focused on the number of productivity of its "domesticated" subjects as a shepherd might husband his flock or a farmer tend his crops. (Scott 2017: 151)

This, we may venture, is in line with colonial genocide studies that locates three non-deterministic phases: (1) initial confrontation (or invasion); (2) carceration period (displacement/resettlement); and (3) assimilation that underlies state formation. The assimilation period, however, tends to rely on enchantment, allure and self-management as binding principles (see Dunlap 2018a). While enchantment and allure might be underplayed in Perlman—likely maintaining that this civilized order was harmful to everyone involved in various degrees and intensities—the individual's struggle with absorption into, and performing, the work of entrails is nowhere better described than through Perlman's (2010 [1983]) deployment of 'armour' and 'masks' to describe the subordination of the human spirit. Indulge this lengthy description:

> The ditch-fixing is something he [sic] takes on to keep from being slaughtered; it is something he merely wears, like a heavy armor or an ugly mask. He knows he will throw off the armor as soon as the Ensi's [overseer] back is turned.
>
> But the tragedy of it is that the longer he wears the armor the less able he is to remove it. The armor sticks to his body. The mask becomes glued to his face. Attempts to remove the mask becomes increasingly painful, for the skin tends to come off with it. There's still a human face below the mask, just as there's still a potentially free body below the armor, but merely airing them takes almost superhuman effort.
>
> And as if all this weren't bad enough, something starts to happen to the individual's inner life, his ecstasy. This starts to dry up. Just as the former community's living spirits shriveled and died when they were confined to the Temple, so the individual's living spirit shrivels and dies inside the armor. His spirit can breathe in a closed jar no better than the gods could. It suffocates. And as the Life inside him shrivels it leaves a growing vacuum. The yawning abyss is filled as quickly as it empties, but not by ecstasy, not by living spirits. The empty space is filled with springs and wheels, with dead things, with leviathan's substance. (Perlman 2010 [1983]: 37–8)

Echoing the concerns of Taussig (1980: 28)—'it appears so natural that the issue of dominance rarely arises'—Perlman here is describing the process of social death and conversion by the violence and phantasms of the

Leviathan. Social death is the hollowing out of spirits (see Card 2003; Short 2016), 'inner life' and 'ecstasy' to reconfigure it—often with the ideology of technological progress via the commodity and consumerism—to create the gears and entrails to propel the Leviathanic apparatus. While only implicitly stated in Perlman, we may say that the 'armour' and 'masks' are features of adult human entrails, not children, whose imaginations are still unsubordinated—or at least still in early phases of domestication. While disregarding children's play, songs and rhymes that invoke another world,[8] the Leviathan is eating this world, taking form through industrial and cybernetic infrastructure as the threat of technological singularity looms. The Devil, then, is endowed with a world and the means to create a body for itself. The Worldeater arises.[9]

Colonizing the Earth: The Virus and Its *Technique*

The Worldeater(s) consumes, reconfigures and converts. The civilized, but more recently techno-capitalist-industrial system has sought to collate and control human and non-human 'populations' and/or 'resources',[10] while simultaneously trying to spread its values, planning geometry, market relationships and, overall, relationship of control across the world, in viral fashion. This metamorphosing worm-octopus-Leviathanic beast emblematic of industrial civilization seeks to touch, influence and subsume vitality. This monstrosity has always, and continuously, been articulated by those hostile to the Worldeater as part of their work of resistance: 'To name these horrors', writes David McNally (2012: 114), 'is also to perform a counter-magic to the sorcery of capital. For capital's great powers of illusion lie in the way it invisibilises its own monstrous formation'. Such counter-magic is found among the Indigenous, peasant and neophyte proletarians described by Taussig. It is found in Marx's abundant images of vampires and gravediggers (see Baldick 1987). And it is found in striking contemporary form among the hostile Zapatistas (2016) who refer to this Leviathanic beast as a mythological 'Hydra' that subsists on death and destruction—war (see Maldonado-Torres 2008)—whose

[8] Consider Simon Springer's (2016b) work on the geography of childhood.

[9] We can imagine this Worldeater taking reference in films like Star Wars in the image of the Death Star to name one among many, but at issue here, especially with the continuing interest with space exploration, is that the Worldeater will spread and begin consuming other planets in the universe, a trajectory in process (see Jakhu et al. 2017).

[10] Creating new words and language to reinforce a particular vision and Stirnarian ghost.

many heads re-grow (often stronger) when they are severed, and is firmly rooted in the privatization of the commons (EZLN 2016: 179, 244–56). The mechanisms mentioned above—hierarchies, symbolic culture, patriarchy and divisions of labor/specialization—remain crucial discursive and practical technologies long identified by green and anti-civilization anarchists (see Zerzan 1988; GA 2012) and, at least to a certain degree, autonomists (Federici 2009 [2004]). While Dunlap (2014, 2018c) asserts that the colony model is the state, colonial genocide scholar Lorenzo Veracini (2014) has found it heuristically useful to discuss 'colonialism as a viral form', specifically relating colonialism to a virus and settler colonialism to bacteria. 'Viruses first attach to a host cell and then penetrate it. They do not have, however, their own metabolism and require a host cell to replicate', says Veracini (2014: 618–9):

> Some viruses are "virulent" and cause disease; others, on the other hand, are latent and allow the host cell to function normally. At times this normalcy is only temporary; indeed, viral infections are characterized by more or less prolonged incubation periods.

Offering excellent detail, Veracini (2014) admits that this type of language risks naturalizing colonial process. Indeed, what could be more natural than viruses that wipe out and kill living organisms? This perspective, however, was taken up earlier to describe the process of 'development' by Majid Rahnema (1997: 116) who refers to development as 'another variety of AIDS' or the 'socio-cultural' type known as 'AIDS II'. The difference between colonization and development for Majid is that colonization is the invasion of an external master-slave relationship, prioritizing coercive intervention, while development, on the other hand, tends to colonize from within—acting as the 'intimate enemy'—changing convivial societies into market (and export) oriented societies organized around money to meet social needs. 'To go back to the metaphoric language of AIDS', says Rahnema (1997: 119), 'it is only when the invader "docks" with the cell and penetrates the membrane that it strips off its protective shell and takes up permanent residence. From now on, the body's own cell becomes the invader'. While Rahnema (1997), following Ivan Illich (1970), discusses the importance of fabricating 'addictive needs', schooling and the 'dis-valuation' of indigenous knowledges, he is ultimately describing the process of social death—the hollowing out of cultural values and subjectivities—to organize the self-management of the techno-capitalist industrial system. The theme and area of social death remains a central theoretical thread throughout these authors (see Card 2003; Short 2016; Dunlap 2018a), even if they do not directly use this term. Both Veracini and Rahnema watch as the world is being consumed by a virus, but equally possible is that these viruses are really a spirit or series of spirits that comprise the accumulation of bodies and ecosystems to give the Worldeater form.

This networked and ecological spread of techno-industrial development is exactly what was theorized by the Spanish engineer Ildefonso Cerdá (1867) in his four-volume *General Theory of Urbanization* (*Teoría general de la urbanización*). Debuting the same year as Marx's *Das Kapital*, the four volumes respond, according Ross Exo Adams (2014: 14), to 'the crisis of the world'—the crisis of industrial development. Adams (2014: 14) describes Cerdá's vision as solving this crisis

> through modern infrastructure and joined together in a new kind of universal network of peace: material circulation to conspire against the political stasis of the state. He would accomplish this through the power of what he called *urbanización*—a term he coined 150 years ago, giving life to a concept and laying out the framework for a new, concrete order of modern life.

Dripping with patriarchal overtones, Cerdá in nineteenth-century utopian fashion theorized urbanization as the process of consuming and interconnecting the earth. A vision frustrated at the time, yet a vision that we nevertheless see coming to fruition with the 'permanent imperative to expand a formless, beige grid of habitation and circulation across the surface of the globe' (Adams 2014: 26, see Fig. 2.2). A grid we see manifesting through the proliferation of industrial-infrastructural corridors (see Hildyard 2016), such as India's Bharatmala and Sagarmala projects integrating the biomass of the country through expressways, while tying it to external markets through seaports (see Seetharaman and Sharma 2018), but also transnational supergrids like Plan Puebla-Panama/Mesoamerica Project (see Call 2002), Desertec between Europe and North Africa (see Sarant 2015) and China's Belt & Road Initiative of intercontinental or even global scale.[11] These supergrids connect continents through roads, pipelines, energy infrastructure and megaproject and conservation initiatives (e.g. ecotourism, mineral, hydrocarbon and renewable energy extrac-

[11] See online 'Belt and Road portal' https://eng.yidaiyilu.gov.cn/.

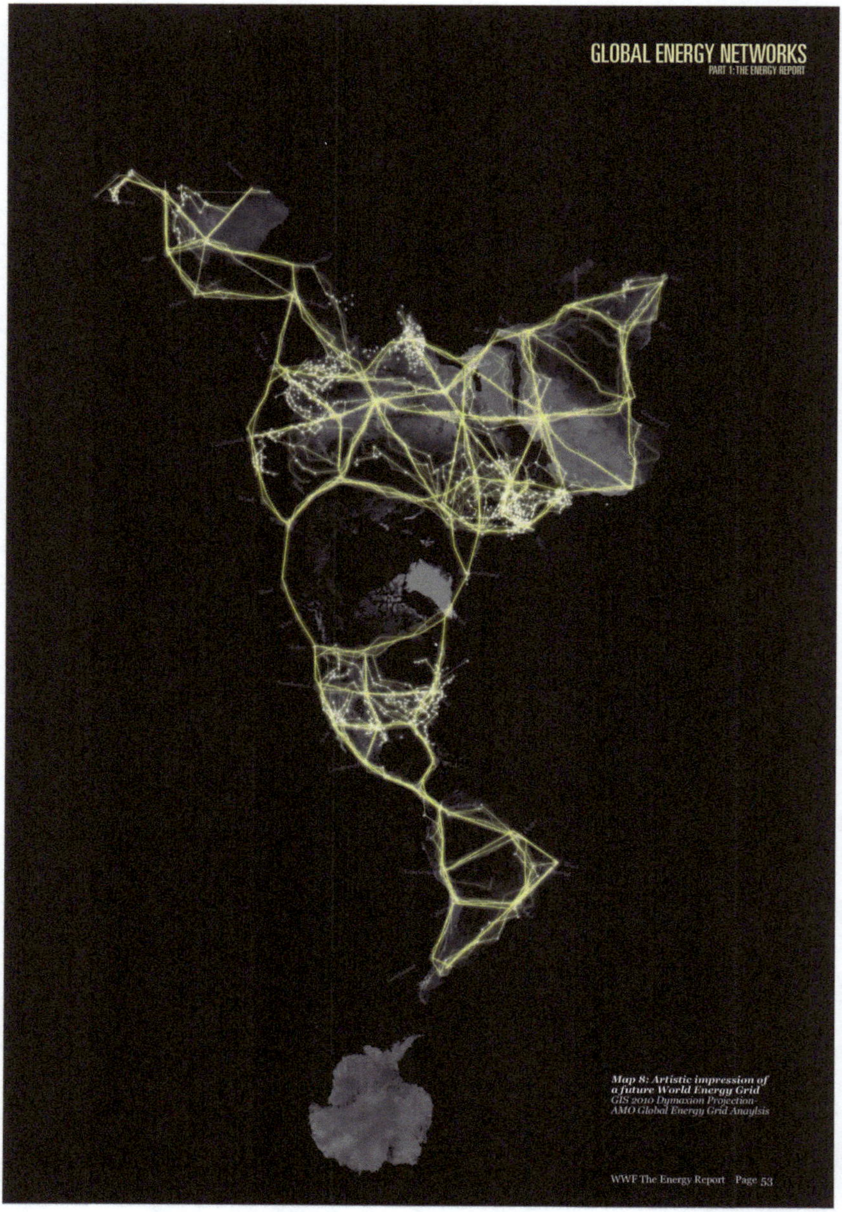

Fig. 2.2 Map of 'world energy regions' and 'world energy grid' (Source: The Energy Report by Metropolitan Office of Architecture (AMO))

tion) that is reconfiguring space and socio-ecological relationships. Stunning are also the ways the grid manifests as it levels the earth's living forests, replacing these with oil palm plantations and mining facilities, cell phone towers or mega highways ensuring connectivity and speed for techno-capitalist progress, but more immediately for our work and consumer convenience. The grid stretches ever farther into remaining rain and old growth forests, bringing with them their vital arteries of road, pipe and power line logistics, converting natural rivers into arteries of logistics while poisoning or displacing fish and other lifeforms. While this vision of dominating the earth through infrastructure and urbanizing processes has been slow, though currently increasing and becoming omnipresent, Adam's (2014: 26) contends it is 'Cerdá's true legacy'.

Spreading this infrastructural control has required the organization of people through state formation. The civilizing virus or Leviathanic beast historically spreads by means of warfare and trade. This process is about transforming horizontal and stateless societies into hierarchical ones (Clastres 1989 [1974]; Scott 2009; Graeber 2012). 'A society needs to be accustomed to having leaders for a foreign power to effectively be able to appoint puppet rulers', explains Peter Gelderloos (2017: 22), continuing:

> Those societies that already have traditional forms of hierarchy, though these might not be enough to qualify them for statehood, are more easily forced into statist logic. If a stateless people has no local, traditional forms of hierarchy that can be exploited by a colonizing state, or if the local leadership—the potential chiefs—cleave to the popular values of anti-authoritarianism and autonomy, a colonizing state has very few possibilities to expand its control. It can either attempt a policy of genocide through extermination or resettlement, or accept the autonomy for the stateless society, at most demanding tribute, a sort of blackmail by which the stateless people produces trade goods to buy reprieve from punitive military actions.

If a colonial/statist power cannot subjugate an indigenous or rebellious people through direct force, then they have to develop more sophisticated methods of wearing them down, having these communities be touched, ingest and digesting the values of invading societies—creating situations where they engage in self-acculturation to the dominant values of the Worldeater to avoid military invasion.

When warfare can be resisted, trade and divide and conquer strategies become the preferred method. Diplomacy and legitimation actions develop simultaneously, alongside forms of trade and production, which is why Perlman (2010 [1983]: 29) in a romantic prose reminds us: 'In the state of nature, trade is something people do to their enemies'. In situations of conquest, or colonial cold wars, the idea is to stress people, divide tribes—often through pre-existing tensions—and to enchant them with cloths, weapons and other fruits of civilized development. Resonating with Veracini (2014) and Rahnema (1997), when colonial powers are defeated and/or kept in place from invading certain areas, they begin by territorializing space with 'factories' and 'trading posts'—the early articulations of 'reservations' (concentration camps) and model villages (see also Dunlap and Fairhead 2014). Demarcating their space and zone of influence, Gelderloos (2017: 24) describes the established 'factories' as 'fortified port settlements with warehouses, barracks, administrative centers, courthouses, prisons and other buildings that fulfilled the joint function of commerce and state-building'. Emblematic of the colony model (see Dunlap 2018a), the trading post—like the factory—in North America was accepted 'to minimize the influence of the encroaching state while availing themselves of the benefits of trade' (Gelderloos 2017: 26). Yet, it was not long before the trading post served as model for the 'reservation' and native collaborators as the perfect functionaries for governing native territories under the United States. It is through these settlements that the Worldeater crawls and its socio-cultural viruses spread.

The unifying logic—or disease—of this spirit is *technique* (Ellul 1964 [1954]). Similar, and more recent, to the notion of technique is James Scott's (1998) 'high-modernist ideology'. Scott (1998: 4) defines it as:

> [A] strong, one might even say muscle-bound, version of the self-confidence about scientific and technical progress, the expansion of production, the growing satisfaction of human needs, the mastery of nature (including human nature), and, above all, the rational design of social order commensurate with the scientific understanding of natural laws.

In line with humans forming the entrails of the Leviathan, we could say the ideologues—or specialists—articulating a high-modernist ideology, such as modernist planners, trans-humanists and the general blind faith into (modern) scientific epistemology comprise the muscular fibers of the beast. 'Technological Wonder proceeds to generate outsiders inside its own entrails', Perlman (2010 [1983]: 300) reminds us, 'to expunge human zeks [captives] and replace them with machines, with things made of its own substance'. This ideology emblematic of 'mechanical philosophy', 'anti-nature' and its combination of more advanced cybernetic-based computational theories (see TIQQUN 2011 [2001]; Galloway 2014) forms the governing, maybe even enslaving, ideology of the Worldeater. This manifests in techno-addiction and workaholism (see Porter and Kakabadse 2006), which are self-reinforcing and instituted through work.

This technological subsumption was theorized and predicted by Jacques Ellul. While Scott's high-modernist ideology is instructive, Ellul's (1964 [1954]) *technique* was precisely concerned with humans becoming appendages of technological progress. Criticizing Lefebvre and neo-utopists like Cerdá, Ellul (1980: 19) tells us:

> I know the glorious arguments about how utopia will open up the imagination and grant us a marvelous freedom. But precisely and concretely, I believe that this trend is actually a "new ruse of the devil" to trick us into entering the megamachine. We must remember that all utopians of the past, without a single exception, have presented society exactly as a megamachine. Each utopia has been an exact repetition of an ideal organization, a perfect conjunction between the various parts of the social body. Utopia presents a flawless totalitarian society.

Ellul, an enormous influence on Perlman, acknowledges this formation of a megamachine and its Devil spirit, viewing this spirit and utopia's, as centered around technique. '[T]echnique is nothing more than means and the ensemble of means', describes Ellul (1964: 19, xxvi) who proceeds to define technique as 'the totality of methods rationally arrived at and having absolute efficiency (for a given stage of development) in every field of human activity'. While the techniques change over time, the ethos of *technique*—the spirit of Worldeater(s)—remains constant. Resonating with Gelderloos' (2017) later observation of colonial power's territorialization with factory and reservations, Ellul (1964: 224) reflects on the enchantment with Nazi technique and its post-World War II articulation:

> The technical system of concentration camps has proved so efficient and satisfactory to the state that it is increasingly being incorporated into our society. It no longer represents the activity of aberrant dictators, but rather the activity of every good administrator.

Complementing Stirner's (2017 [1845]: 204) early theorization of 'prison society', Ellul draws continuity between, not only the North American reservation, German and English colonial internment camps as coalesced into Nazi concentration camps (see Arendt 1962 [1951]), but how the technique of scientific violence and population management would only become normalized in planning. We see an intensification of technique and ultimately the modality by which nature is paved over and suffocated by industrial infrastructure and humans are corralled and managed in market societies, which takes on previously unimaginable—science fiction—levels with the rise of digital technologies (see Graham 2011). Moreover, this resonates with Sale's epigraph above and anarchist insights of hierarchy and domination. Springer et al. (2019) and green anarchists in general have demonstrated clearly that once you dominate, domesticate and cage non-human life—flora and fauna—it is not long before these patters take hold, where humans do this to each other often through insecure theories of difference, 'race' and biology (see Hinton 2002). When factory farms are permitted to slaughter animals for food, the proliferation of sweatshops and (privatized) prison labor will not be far behind (see White 2017). Meanwhile, the possibilities of eating humans, as depicted in the film *Soylent Green* (1973), will always loom until these practices are abolished. In the utopian dreams associated with mechanical philosophers and utilitarians such as Bentham's Panopticon, Haussmann's planning, Cerdá's urbanization, Le Corbusier's architecture—among other contemporaries and disciples—we witness the birth of their monstrous offspring: A shift in colonial/statist production that journalists Robert Davis and Mark Zannis (1973) would call the genocide machine—another theory underpinning the Worldeater.

Welcome to the (Genocide) Machine

Colonial Genocide scholars A.D. Moses (2002) and Damian Short (2010, 2016) have acknowledged the value of the theory of the genocide machine, if not advancing and updating its ideas. Like nearly all the authors referenced above, Davis and Zannis (1973: 13) recognize how the process of industrial development is only 'becoming more dehumanized' with the 'advance of technology'. Philosopher Grégoire Chamayou (2015 [2013]) would call this progress the '[i]ndustrial production of compartmentalized psyches, immunized against any possibility of reflecting upon their own violence', which again reverberates with Taussig's (1980) theme of normalizing the violence of market society so the Devil's work of extraction—and possibly bodily realization—could proceed. The theory of the genocide machine contends, in the words of Davis and Zannis (1973: 33), that 'colonialism has absorbed a genocide through detached technological means, rather than by inflamed human emotions of fear and hatred. The rhetoric has been cooled and the action automated'. We should stress here that the classism, racism and sexism inherent to industrial society and colonialism was also absorbed and embedded into these technological systems, yet the'[a]cts of genocide'—specifically the process of social death—'become [increasingly] polite and clinical' (Davis and Zannis 1973: 33). An extension of traditional colonialist genocide, the Genocide Machine represents developmental progress—an economization of operations—geared toward harnessing life forces, as opposed to exterminating them to control land and assert supremacy. The Genocide Machine weds biopolitics and necropolitics to the cause of techno-capitalist progress (Foucault 1998 [1978]; Mbembé 2003), embodying a tension toward the total economization of resources to expand political economy. This necessitates creating obedient and self-managed regimes to facilitate industrial production and consumption. Violence becomes strategic and 'smart', powerful international corporatist actors 'murder only when they are forced to by resistance' (Davis and Zannis 1973: 176; see also Dunlap 2018a)—or as a private security contractor, 'Jim', discussing mining in Peru explained: 'when a problem has been created that cannot be controlled. When they cannot control the situation with money' (Dunlap 2019a: 20), the last resort is to 'disappear' anti-mining agitators.

The Worldeater—industrial techno-capitalist civilization—has learned to better economize, to better utilize and harness the resources of the earth in its quest to create a body and to materialize and, we suggest, continue consuming other planets—interstellar colonization (see Jakhu et al. 2017). Resonating with Rahnema Majid's, (1997) conception of development, the violence is internalized and normalized—people are possessed—as political economy is organized to promote the self-management that advances technological progress and consumerism, both preying on the fears—control/insecurity issues—that market society has only exacerbated (see Alexander 2008). Part of economizing requires maintaining legitimacy and employing humanitarianism for such a purpose. Following Eyal Weizman (2011), we call this the 'lesser evil technique'. Weizman (2011: 10) recognizes that that 'less brutal measures are also those that may be more easily naturalized, accepted and tolerated—and hence more frequently used, with the result that a greater evil may be reached cumulatively'. The Worldeater is economizing through humanitarianism; meanwhile states applaud white supremacy and neo-fascism(s) to accelerate regimes of extractivism. The good/bad cop of total extractivism is forming (see Chap. 4), where the green economy articulates a lesser evil—justifying ecological modernization and growth with dystopic implications (see the conclusion in Dunlap 2019b). Meanwhile, conventional extractivism and authoritarianism—the 'bad cop'—are blooming under figures like Trump, Bolsonaro and others leaders across the world who are backed by extractive industries, and their megalomania and thrust for power epitomizes the fear for control and meaning-making that is leading to the end of humans and non-humans as we know it. In sum, the Worldeater is techno-capitalist progress and it is a spirit possessing us, teaching us to sacrifice ourselves and our environments at its altar—the altar of progress that will give it material form, a body.

How Worldeating Progresses: The Rest of the Book

This chapter offers the concept of the Worldeater, as a tension embodied in technique, capitalism and, by extension, the driving force of extractivism. While we see great value in understanding the spirits of Worldeater(s) as a mode of resistance, and supportive of land-based and pre-capitalist cultures, we still offer a genealogy uniting various authors, all discussing the same phenomenon of capitalist subsumption, genocide and planetary change driven by industrial development.

The Worldeater framing offers important theoretical insights. First, it opens up limited ontological and epistemological readings, by acknowledging mythological and psycho-spiritual warfare embedded in the Worldeater.[12] Second, as mentioned, it conceptualizes a virus and a possession that has the potential to link individual habits (consumerism, insecurity/control issues, political submission, etc.) with environmental processes (climate change, deforestation, ocean acidification, pollution/industrial wastes, etc.). The Worldeater is not absolute, but its contagion is variegated, shifting and operating on differential intensities within people and institutions. The Worldeater manages people's 'rational best interest' to discourage revolt and entice assimilation, meanwhile organizing a material and interspecies political division of labor centered on class, sex, 'race', and real or imagined differences more generally. The ideology of techno-capitalist progress (of which Scott's (1998) 'high-modernist ideology' is a

[12] This could refer to psychological war and media operations or, possibly, origins in shamanistic warfare; the question remains open for exploration.

key exemplar) transcends various political regimes, incorporates the relations of capital—class, exploitation, dispossession—and is at the heart of the Worldeater and its dispersion of violence, allure and indifference amongst its converted participants or, in Perlman's (2010 [1983]: 29) words, 'entrails'. Third, as indicated, Perlman (2010 [1983]) and others (Zerzan 1988, 2005 [1999]; Landstreicher 2009)[13] demonstrate that the evolutionary rooting of the Worldeater is older than capitalism. The onset of ancient civilization birthed the spirit and rudiments of the Worldeater that evolved through capitalism into the present. Fourth, the Worldeater thrives on human (existential) insecurity. This resonates with the 'genocide machine' (Davis and Zannis 1973) that 'is characterized by a pervasive, repressed fear that corrodes the values and sanity of subject peoples and colonial powers alike'. This 'repressed fear' and value corrosion is embedded in socio-technical systems organized around human and non-human hierarchies/speciesism (human supremacy/inferiority complex); patriarchy/gender; and (extreme) divisions of labor/specialization that domesticates people to the imperative of the Worldeater. This describes the roots within the structure of conquest, which, following Perlman (2010 [1983]), works toward *entrails-making*. The planet, we believe, is confronted with the Worldeater epoch, not the so-called Anthropocene. This, we contend, better resituates the complexity of the problems facing humanity—the actors and institutions most responsible—while repositioning peoples agency in an unbalanced global habitat. Fifth and relatedly, the longer an individual is subsumed—in whole or in part—by the entrails of the Worldeater, it is more difficult to see, think, feel and speak outside its logic, perspective and terminology. Perlman (2010 [1983]: 54) writes: 'All beings not encased in its entrails, whether people or animals or trees, are its enemy'—an enemy to be captured, captivated or erased physically or culturally.

We have highlighted various and related roots for reconsidering the foundations of social, economic and ecological crises. A central point being, echoing Gelderloos (2017), the importance of 'chaotic organization'—as opposed to decentralized. Do not forget, asserts Paul Virilio (2008 [1983]: 111), 'that this desire for self-management coexists with a desire for hyper-centralization, which is the result of technology'. Chaotic organization should consider this as well, but the point is that there should be a permanent tension toward the imposition and domination of *technique*—the construction of mechanical/cybernetic systems—that necessi-

[13] See also Green Anarchy Magazine (2002–2009), Return Fire (2013–Present), and Black Seed (2014–Present).

tate and facilitate planetary resource extraction activities, but most importantly, impoverished socio-ecological relationships.

With this theoretical framing and provocation in place, the subsequent book chapters take on a more normalized academic prose. Now that we have framed the problem, Chap. 3 examines the commonalities and differences between the two sub-disciplines political ecology and critical agrarian studies. This disciplinary—internal academic conversation—serves three purposes. First, we want these disciplines to take up the concept of the Worldeater—expand upon it, debate it, put it to work on empirical cases, criticize it. Second, we believe these subfields have the strongest intellectual tools and interests in charting the progress of the Worldeater(s). Third, we want to offer a genealogy to be clear about the strengths and weaknesses of these disciplines, to acknowledge where they come from and where they are going, which serves more academic purposes than the conversation around the Worldeater. Chap. 4 then moves to glance at the claws, teeth and venom of the Worldeater—in other words, the repressive violence employed to facilitate and maintain extraction. This involves examining the relationships between resource control and militarization, to support the what is presented here (in Chap. 2) and to see the evolution and growth of the Leviathanic beast in how it secures resources. Following this, Chap. 5 then dissects the blurring of conventional and 'green' forms of natural resource extraction, tracing the contours of the systemic colonizing grid. This outline foregrounds how Perlman (2010 [1983]: 5) described his compass as he set out on his pioneering journey into the abyss of the Worldeater:

> It is my aim to speak of the beast's body. For it does have a body, a monstrous body, a body that has become more powerful than the Biosphere. It may be a body without any life of its own. It may be a dead thing, a huge cadaver. It may move its slow thighs only when living beings inhabit it. Nevertheless, its body is what does the wrecking.

This is what is at issue when we examine academic sub-disciplines, coercive violence and the merging of conventional and green extractivism in the subsequent chapters. Mapping the organization and tension of total extractivism—as an imperative and threatening force at the current conjuncture—the chapters offers the reader glimpses into how the world, as we know it, is being transformed, converted and consumed.

Bibliography

Adams, R. E. (2014). Natura Urbans, Natura Urbanata: Ecological Urbanism, Circulation, and the Immunization of Nature. *Environment and Planning D: Society and Space, 32*, 12–29.

Alexander, B. K. (2008). *The Globalization of Addiction: A Study in Poverty of the Spirit*. New York: Oxford University Press.

Arendt, H. (1962 [1951]). *The Origins of Totalitarianism*. New York: The World Publishing Company.

Arrighi, G. (1994). *The Long Twentieth Century: Money, Power, and the Origins of Our Times*. New York: Verso.

Baldick, C. (1987). *In Frankenstein's Shadow: Myth, Monstrosity, and Nineteenth-Century Writing*. Oxford: Clarendon Press.

Black Seed. (2014–Present). *Black Seed: A Green Anarchist Publication*. Retrieved from http://blackseed.anarchyplanet.org/.

Bookchin, M. (1982). *The Ecology of Freedom: The Emergence and Dissolution of Hierarchy*. Palo Alto, CA: Cheshire Books.

Call, W. (2002). Plan Puebla-Panama. *NACLA Report on the Americas, 35*, 24.

Card, C. (2003). Genocide and Social Death. *Hypatia, 18*, 63–79.

Cerdá, I. (1867). *Teoría general de la urbanización, y aplicación de sus principios y doctrinas a la reforma y ensanche de Barcelona*. Madrid: Imprenta Española.

Chamayou, G. (2015 [2013]). *A Theory of the Drone*. New York and London: The New Press.

Clark, J., & Martin, C. (2013). *Anarchy, Geography, Modernity: Selected Writings of Elisée Reclus*. Oakland: PM Press.

Clastres, P. (1989 [1974]). *Society Against the State: Essays in Political Anthropology*. New York: Zone Books.

Clastres, P., Hurley, R., & Stein, A. (1977). *Society Against the State*. Oxford: Blackwell.

Davis, R., & Zannis, M. (1973). *The Genocide Machine in Canada*. Montreal, CA: Black Rose Books.

Derrida, J. (2012 [1993]). *Specters of Marx: The State of the Debt, the Work of Mourning and the New International*. London: Routledge.

Duffield, M. (2011). Total War as Environmental Terror: Linking Liberalism, Resilience, and the Bunker. *The South Atlantic Quarterly, 110*, 757–769.

Dugger, W. (1989). *Corporate Hegemony*. Westport, CT: Greenwood Press.

Dunlap, A. (2014). Permanent War: Grids, Boomerangs, and Counterinsurgency. *Anarchist Studies, 22*, 55–79.

Dunlap, A. (2018a). The 'Solution' Is Now the 'Problem': Wind Energy, Colonization and the 'Genocide-Ecocide Nexus' in the Isthmus of Tehuantepec, Oaxaca. *The International Journal of Human Rights, 42*, 550–573.

Dunlap, A. (2018b). "A Bureaucratic Trap": Free, Prior and Informed Consent (FPIC) and Wind Energy Development in Juchitán, Mexico. *Capitalism Nature Socialism, 29*, 88–108.

Dunlap, A. (2018c). *Reconsidering the Logistics of Autonomy: Ecological Autonomy, Self-Defense and the Polícia Comunitaria in Álvaro Obregón, Mexico*. Retrieved from https://www.tni.org/files/article-downloads/erpi_cp_8_dunlap.pdf.

Dunlap, A. (2019a). 'Agro sí, mina NO!' The Tía Maria Copper Mine, State Terrorism and Social War by Every Means in the Tambo Valley, Peru. *Political Geography, 71*, 10–25.

Dunlap, A. (2019b). *Renewing Destruction: Wind Energy Development in Oaxaca, Mexico*. London: Rowman & Littlefield.

Dunlap, A., & Fairhead, J. (2014). The Militarisation and Marketisation of Nature: An Alternative Lens to 'Climate-Conflict'. *Geopolitics, 19*, 937–961.

Ellul, J. (1964 [1954]). *The Technological Society*. New York: Vintage Books.

Ellul, J. (1980). *The Technological System*. New York: Continuum.

EZLN. (2016). *Critical Thought in the Face of the Capitalist Hydra: I: Contributions by the Sixth Commission of the EZLN*. Brisbane: Paper Boat Press.

Federici, S. (2009 [2004]). *Caliban and the Witch: Women, the Body and Primitive Accumulation*. New York: Autonomedia.

Fitzpatrick, B. (2018). *Corrosive Consciousness*. Jacksonville, FL: Enemy Combatant.

Foucault, M. (1998 [1978]). *The Will to Knowledge: The History of Sexuality: 1*. London: Penguin Books.

Foucault, M. (2003 [1997]). *"Society Must Be Defended": Lectures at the College de France 1975–1976*. New York: Picador.

Fukuoka, M. (2010 [1978]). *The One-Straw Revolution: An Introduction to Natural Farming*. New York: New York Review of Books.

GA (Green Anarchy). (2002–2009). *Green Anarchy: An Anti-Civilization Journal of Theory and Action*. Retrieved from http://greenanarchy.anarchyplanet.org/.

GA. (2005). *What Is Green Anarchy? An Introduction to Anti-Civilization Anarchist Thought and Practice*. Retrieved from https://theanarchistlibrary.org/library/anonymous-what-is-green-anarchy.

GA. (2012). *Uncivilized: The Best of Green Anarchy*. Berkeley, CA: Ardent Press.

Galloway, A. R. (2014). The Cybernetic Hypothesis. *Differences, 25*, 107–131.

Gelderloos, P. (2017). *Worshiping Power: An Anarchist View of Early State Formation*. Oakland, CA: AK Press.

Gorrion, A. (2013). Anarchy in World Systems. *The Anvil Review*. Retrieved from http://theanvilreview.org/print/anarchy-in-world-systems/.

Graeber, D. (2012). *Debt: The First 5,000 Years*. London: Melville House.

Graham, S. (2011). *Cities Under Siege: The New Military Urbanism*. New York: Verso.

Han, C. (2017). *Psycho-Politics*. London: Verso.

Hedegaard, H., Curtin, S. C., & Warner, M. (2018). *Suicide Rates in the United States Continue to Increase*. Retrieved from https://www.cdc.gov/nchs/products/databriefs/db309.htm.

Hildyard, N. (2016). *Licensed Larceny: Infrastructure, Financial Extraction and the Global South*. Manchester: Manchester University Press.

Hinton, A. L. (2002). *Annihilating Difference: The Anthropology of Genocide*. Berkeley: University of California Press.

Illich, I. D. (1970). *Planned Poverty: The End Result of Technical Assistance. Celebration of Awareness: A Call for Institutional Revolution*. London: Marion Boyars.

Jakhu, R. S., et al. (2017). *Space Mining and Its Regulation*. Chichester: Springer.

Jensen, D. (2016). *The Myth of Human Supremacy*. New York and Oakland, CA: Seven Stories Press.

Landstreicher, W. (2009). *Willful Disobedience*. Berkeley, CA: Ardent Press.

Maldonado-Torres, N. (2008). *Against War*. Durham, NC: Duke University Press.

Mbembé, A. (2003). Necropolitics. *Public Culture, 15*, 11–40.

McNally, D. (2012). *Monsters of the Market: Zombies, Vampires, and Global Capitalism*. Chicago: Haymarket Books.

Merchant, C. (1980). *The Death of Nature: Women, Ecology, and the Scientific Revolution*. New York: Harper & Row.

Mies, M., & Shiva, V. (2014 [1993]). *Ecofeminism*. London: Zed Books.

Moses, A. D. (2002). Conceptual Blockages and Definitional Dilemmas in the 'Racial Century': Genocides of Indigenous Peoples and the Holocaust. *Patterns of Prejudice, 36*, 7–36.

Naess, A. (1973). The Shallow and the Deep, Long-Range Ecology Movement. A Summary. *Inquiry, 16*(1–4), 95–100.

OECD. (2017). *Health at a Glance: 2017 OECD Indicators*. Retrieved from https://www.oecd-ilibrary.org/social-issues-migration-health/health-at-a-glance-2017/antidepressant-drugs-consumption-2000-and-2015-or-nearest-year_health_glance-2017-graph181-en.

Perlman, F. (2010 [1983]). *Against His-Story, Against Leviathan*. Detroit: Red and Black Press.

Platt, J. (2013). Why Winona LaDuke Is fighting for Food Sovereignty. Retrieved from http://www.mnn.com/leaderboard/stories/why-winona-laduke-isfighting-for-food-sovereignty.

Polanyi, K. (2001 [1944]). *The Great Transformation: The Political and Economic Origins of Our Time*. Boston: Beacon Press.

Porter, G., & Kakabadse, N. K. (2006). HRM Perspectives on Addiction to Technology and Work. *Journal of Management Development, 25*, 535–560.

Pratt, L. A., Brody, D. J., & Gu, Q. (2017). *Antidepressant Use Among Persons Aged 12 and Over: United States, 2011–2014*. NCHS Data Brief. Number 283. National Center for Health Statistics Brief 283. Retrieved from https://www.cdc.gov/nchs/data/databriefs/db283.pdf.

Rahnema, M. (1997). Development and the People's Immune System: The Story of Another Variety of AIDS. In M. Rahnema & V. Bawtree (Eds.), *The Post-Development Reader* (pp. 111–129). London: Zed Books.

Reclus, E. (2013 [1905]). *Anarchy, Geography, Modernity: Selected Writings of Elisée Reclus.* Oakland, CA: PM Press.

RF (Return Fire). (2013–Present). *Return Fire: Anti-Authority-Daily Revolt-Individual Will-De-Civilisation.* Retrieved from https://325.nostate.net/distro/.

RF. (2016) *Colonisation.* Retrieved from https://en-contrainfo.espiv.net/files/2016/06/Colonisationreturn-fire-vol3.pdf.

Sale, K. (1991 [1985]). *Dwellers in the Land: The Bioregional Vision.* Athens: University of Georgia Press.

Sarant, L. (2015). *Desertec: An Aborted Project or Just a Change of Direction?* Retrieved from https://www.natureasia.com/en/nmiddleeast/article/10.1038/nmiddleeast.2015.4.

Sayer, D. (1987). *The Violence of Abstraction: The Analytical Foundations of Historical Materialism.* New York: Basil Blackwell.

Scott, J. C. (1998). *Seeing Like a State: How Certain Schemes to Improve the Human Condition Have Failed.* New Haven, CT: Yale University Press.

Scott, J. C. (2009). *The Art of Not Being Governed: An Anarchist History of Upland Southeast Asia.* New Haven, CT: Yale University Press.

Scott, J. C. (2017). *Against the Grain: A Deep History of the Earliest States.* New Haven, CT: Yale University Press.

Seetharaman, G., & Sharma, S. N. (2018). Find Out How the Government's Marquee Infrastructure Projects Are Doing. *The Economic Times.* Retrieved from https://economictimes.indiatimes.com/news/economy/infrastructure/find-out-how-the-governments-marquee-infrastructure-projects-are-doing/articleshow/65273342.cms?from=mdr.

Short, D. (2010). Cultural Genocide and Indigenous Peoples: A Sociological Approach. *The International Journal of Human Rights, 14*, 833–848.

Short, D. (2016). *Redefining Genocide: Settler Colonialism, Social Death and Ecocide.* London: Zed Books.

Springer, S. (2016a). *The Anarchist Roots of Geography: Toward Spatial Emancipation.* Minneapolis, MN: University of Minnesota Press.

Springer, S. (2016b). *Learning Through the Soles of Our Feet: Unschooling, Anarchism, and the Geography of Childhood.* London: Rowman & Littlefield.

Springer, S. (2017). The Limits to Marx: David Harvey and the Condition of Postfraternity. *Dialogues in Human Geography, 7*(3), 280–294.

Springer, S., Locret, M., Mateer, J., et al. (2019). *Anarchist Political Ecology.* Oakland, CA: PM Press.

Stirner, M. (2017 [1845]). *The Unique and Its Property.* Berkeley, CA: Ardent Press.

Taussig, M. T. (1980). *The Devil and Commodity Fetishism in South America.* Durham, NC: University of North Carolina Press.

Thoreau, H. D. (1971 [1854]). *Walden.* Princeton, NJ: Princeton University Press.

TIC (The Invisible Committee). (2015). *To Our Friends.* South Pasadena: Semiotext(e).

TIQQUN. (2011 [2001]). The Cybernetic Hypothesis. *TIQQUN* 2. Retrieved from https://theanarchistlibrary.org/library/tiqqun-the-cybernetic-hypothesis.

Veracini, L. (2014). Understanding Colonialism and Settler Colonialism as Distinct Formations. *Interventions, 16*, 615–633.

Virilio, P. (2008 [1983]). *Pure War*. Los Angeles: Semiotext(e).

Weber, M. (1992 [1930]). *The Protestant Ethic and the Spirit of Capitalism*. London and New York: Routledge.

Weizman, E. (2011). *The Least of All Possible Evils: Humanitarian Violence from Arendt to Gaza*. London: Verso.

White, R. (2017). Capitalsim and the Commodification of Animals: The Need for Critical Vegan Praxis. In D. Nibert (Ed.), *Animal Oppression and Capitalism*. Connecticut: Praeger.

Zapatistas. (2016). *Critical Thought in the Face of the Capitalist Hydra I*. Durham: Paper Boat Press.

Zerzan, J. (1988). *Elements of Refusal*. Seattle, WA: C.A.L. PressIPaleo Editions.

Zerzan, J. (2005 [1999]). *Against Civilization: Readings and Reflections*. Port Townsend, WA: Feral House.

CHAPTER 3

Studying the Worldeater(s): Political Ecology and Critical Agrarian Studies and Their Origins, Differences and Convergence

Abstract This chapter discusses critical agrarian studies and political ecology, two of the most central academic fields responsible for charting land control, territorialization and extraction in the service of techno-capitalist development. These academic subfields, we can say, specialize in examining the parts and developmental trends of the Worldeater(s). Through an extensive review of critical agrarian studies and political ecology, this chapter shows forgotten disciplinary roots, under-acknowledged commonalities and important differences leading, nevertheless, to increasing convergence within the subfields. This review allows us to calibrate further our analytical tools for the subsequent inquiry into the 'claws and teeth' as well as remaining developmental form—body—of the Worldeater(s).

Keywords Critical agrarian studies • Political ecology • Land control • Territorialization • Extraction

A. Dunlap, J. Jakobsen, *The Violent Technologies of Extraction*,
https://doi.org/10.1007/978-3-030-26852-7_3

Conceiving techno-capitalist progress as the Worldeater, this chapter examines the critical research engaged in charting land control, territorialization and extraction to feed this trajectory of development. While framed in this way, this is a disciplinary discussion to chart the different paths, striking similarities and convergences as well as forgotten influences of political ecology and critical agrarian studies, as we believe these subfields have the strongest intellectual tools and interests in charting the progress of the Worldeater(s). In and of itself, this disciplinary discussion has rarely been attempted at length, precluding us from seeing some of the crucial ways that these sub-disciplines/fields allow for studying the 'claws and teeth' as well as remaining 'body' of the Worldeater(s). We believe these subfields have the strongest intellectual tools and interests in charting the progress of the Worldeater(s). It is from this place that we offer a genealogy to accentuate the strengths and weaknesses of these disciplines, to acknowledge where they come from and where they are going. Finally, it is our hope that these disciplines will take up the concept of the Worldeater—expand upon it, debate it, put it to work on empirical cases, criticize it. Indigenous people, (anti-)anthropologists and land-based people generally (and engaging in the antics of academic writing) retain a special place for developing the roots, weakness and methods to unravel this monster that the Worldeater signifies. Widening the pluriverse of perspectives—by articulating an immanence that decenters the scientism and materialism that dominates—recognizing the Worldeater, paradoxically, performs the necessary intellectual insurrection of the present to make legible the techno-capitalist beast that is swallowing—physically and psycho-socially—the planet. It is our hope that this chapter not only will accentuate the radical anarchistic and pluriversal tendencies already existing within these fields, but also, and prospectively, that scholars in political ecology and critical agrarian studies will muster the courage to actually employ, criticize and develop further the concept of the Worldeater(s) and the trajectory of total extractivism. We would like to see these studies get queer in the widest sense.

Forgotten Siblings: Critical Agrarian Studies (CAS) & Political Ecology

The term critical agrarian studies (CAS) has only recently become an established frame for emerging scholarship arising from peasant studies and, relatedly, agrarian political economy (Edelman and Wolford 2017; Akram-Lodhi 2018).[1] This work focuses on processes of socio-ecological change and differentiation, examining socio-economic relationships of rural people, their productive systems and how they are impacted by land and labor regime changes, or development interventions, as they emanate from 'above'—initiated by governments and corporations—and from 'below'—initiated by local elites, landless workers, farmers and Indigenous populations (Borras and Franco 2013; Edelman and Wolford 2017). Building upon the peasant studies tradition, as we will see, CAS departs from it in a more explicit concern with embedding localized field studies within global processes of capitalist change (Akram-Lodhi 2018). As we can expect from this overall focal interest, CAS has involved significant amounts of research on agroecology (Rosset et al. 2011; Rosset and Altieri 2017); plantation systems (Li 2011; Peluso 2017; Lund 2018); biofuels (Borras et al. 2010); bioprospecting (Neimark 2012; Montenegro de Wit 2017); and conservation (Fairhead et al. 2012). In particular, CAS has become renowned for bourgeoning work on large, medium and small-scale land acquisition or, more accurately, land/green grabbing (Zoomers 2010; Li 2011; White et al. 2012; Borras et al. 2012; Fairhead et al. 2012;

[1] CAS is, as Edelman and Wolford (2017) point out, most visibly institutionalized in the academia through *Journal of Peasant Studies* and *Journal of Agrarian Change*.

Edelman et al. 2013; Aguilar-Støen 2016; Hunsberger et al. 2017; Levien 2018; Vorbrugg 2019). This work, in turn, is interlaced with work on land control (Ribot and Peluso 2003; Peluso and Lund 2011), territorialization (Lund 2016; Rasmussen and Lund 2018) and political reactions 'from below' (Borras and Franco 2013; Hall et al. 2015). The rapid popularity of critical agrarian studies among researchers, activists and even (to a certain extent) policy makers is unprecedented. CAS remains instrumental, in other words, to uncovering the web of total extractivism and the formation of the Worldeater.

Paul Robbins (2012 [2004]: 12) defines political ecology as an 'empirical, research-based exploration to explain linkages in the condition and change of social/environmental systems, with explicit consideration of relations of power'. Meanwhile, Joan Martínez-Alier (2002) views political ecology as a study of ecological distribution conflicts, which is the study of conflicts over access to and control of natural resources, particularly as a source of livelihood, as well as the costs of environmental destruction (see Martínez-Alier and O'Connor 1995; Martínez-Alier 2002; Scheidel et al. 2017). Arturo Escobar (2008: 14) extends this to include cultural distribution conflicts as 'economic crises are ecological crises are cultural crises', which Mario Blaser (2013: 15) takes further with 'the political ontology of "environmental" conflicts', contending that the root of cultural distribution conflicts—'whichever cultural perspective gains the upper hand will determine the access to, use of and relation to "the thing" at stake—is political ontology'. From ecological to ontological distribution conflicts, political ecology has been central to interrogating power-relations in environments in the widest sense (Svarstad et al. 2018). This has led to research focusing on violence and war over natural resources (Peluso and Watts 2001; Le Billon 2001, 2012); conservation and the neo-liberalization of nature (Heynen and Robbins 2005; Brockington and Duffy 2010; Büscher et al. 2012; Sullivan 2006, 2010; Holmes 2014; Fletcher et al. 2018); subsoil extractivism (Gudynas 2009; Bebbington et al. 2008; Bebbington 2012; Lang and Mokrani 2013 [2011]; Bebbington and Bury 2013; Tetreault 2014), the connection between conservation and extraction (Büscher and Davidov 2013; Sullivan 2013b; Duffy 2015; Mckay 2017) as well as social movements and resistance (Martínez-Torres and Rosset 2010; Martinez-Alier et al. 2016).

On such a condensed backdrop, we now proceed to unpack in more detail the formation of the two research fields and their underlying contradictory intertwinement of radical social theory, praxis and military-statist logics. We continue to interrogate differences and convergences between them, highlighting politico-theoretical streams of thought including under-acknowledged anarchist currents. Lastly, we identify one specific avenue for renewed convergence in recent and emerging work that has particular bearing on studies of the imperative of total extractivism.

The Rise of Critical Agrarian Studies

Critical agrarian studies (CAS) is rooted in the peasant studies of the 1960s and 1970s. Conventionally tracing its genealogy to the Marxist tradition(s), overviews of CAS turn to the central position of the 'agrarian question', which is emblematic of Karl Kautsky's (1988 [1899]) questions: 'whether, and how, capital is seizing hold of agriculture, revolutionising it, making old forms of production and property untenable and creating the necessity for new ones' (Akram-Lodhi and Kay 2010a: 179). A foundational element of peasant studies is thus Marx's (1982 [1867]) analysis of capitalism and 'primitive accumulation' with the peasantry interlaced throughout his works (Akram-Lodhi and Kay 2010a; Anderson 2010; Bernstein 2010). This work, recently dubbed 'agrarian Marxism' (Levien et al. 2018), was carried forward by sociologists, economic historians and neo-Marxists, notable among them Karl Polanyi (2001 [1944]), EP Thompson (1991 [1963]), Barrington Moore (1974 [1966]) and Raymond Williams (1973). Thus, peasant studies was profoundly influenced by classical Marxism with Karl Kautsky (1988), V.I. Lenin (1964 [1899]) and N.I Bukharin among others (see Akram-Lodhi and Kay 2010a, 2010b) that maintain a sustained orientation toward questions of class and inequality (Bernstein 2010). Foundational to this field of study was 'the farm', explains Richard Walker (2004: 6), 'because on-farm production appears as something vastly different from modern industry'. In this fundamental farm-centered focus lay the questions that triggered unfolding waves of debates—including the prominent Lenin-Kautsky-Chayanov debates—that have resurfaced ever since (see Bernstein et al. 2018). This abiding farm-centrism no doubt has contributed to the relative lack of acknowledgment of larger trajectories of total extractivism examined in greater depth in Chap. 5.

While it is clear from the above that peasant studies was formed at, and has proceeded in and through, a Marxist conjuncture, there were other influences that should be recognized. Marxism, as Edelman and Wolford (2017: 5) explain, was the first of six additional influences on peasant studies. Following classical Marxism was the 'peasant-centered Marxism' or Maoism, (3) Russian Agrarian economists at the turn of the nineteenth century, (4) economic and agrarian history, (5) heterodox comparative social scientists—notably Eric Wolf (1999 [1969]), Moore (1966) and Scott (1977)—and, finally, (6) scattered works from rural sociology, demography, anthropology and the emerging field of political ecology (see Edelman and Wolford 2017: 5; Akram-Lodhi and Kay 2010a). These rich, cross-disciplinary and, at times, politically contradictory works serve as the foundations of the recent and influential critical agrarian studies.

Another, much less illustrious, genealogical thread to peasant studies is found in military logics. Explaining the rise of peasant studies in the US after the World War II, Marc Edelman and Wendy Wolford (2017: 4) note that 'both the political left and the right viewed the peasantries of Asia, Africa and Latin America as important historical agents'. While this is a 'positive' framing, we want to emphasize that, after the Chinese Revolution, 'the peasant' was perceived as among the top national security threats to US influence and stability (Cullather 2006; Dunlap and Fairhead 2014), which a decade later was complemented by the internal threat of inner cities rioting against racism and political injustice (Light 2003). Both the inner city and the jungle were deemed illegible and rebellious territories harboring insurgency, thus creating a demand for greater knowledge or, more accurately, intelligence on the habits, attitudes and political psychology of the peasant with the intention to tailor policy and civil-military interventions in the countryside (Solovey 2001; Salemink 2003). Based on this notion of threat, the US Department of Defense (DoD) 'from the outset sponsored most social science research on Vietnam and peasant rebellions' (Salemink 2003: 171), funding peasant studies and attempting to position it in the service of the US counterinsurgency apparatus. In the case of Vietnam, 'USAID officials and others began to show interest in the history of tenancy, rents, landlords, [and] taxation in colonial times', writes Oscar Salemink (2003: 183) as land reform could serve as a non-military pacification strategy, filling the income-inequality gap between 'town and country' and improving governmental communication to rural Vietnam. While the United States is apprehensive to implement land reform in any meaningful way (see Copeland 2012; Albertus and Kaplan 2013), it remains an essential political concession and, consequently, policy tool in the arsenal of counterinsurgency. Social science in the service of warfare is not new (see Solovey 2001; Price 2011, 2014), but, what is interesting here is that the peasant rebellion and the US government's responses to them are among the principle architects of peasant studies.

Noticeably, peasant studies were deeply concerned with understanding the land regimes, cultures, socio-ecological relationships, habits, political psychology and actions taken by rural populations. Much of this stemmed from efforts at social engineering socio-statist systems in Europe, understanding developmental transitions, life ways, reasons and methods of revolt. The crossover with anthropological research into human origins, Indigenous societies and—directly and indirectly—advancing colonial control is pronounced (see Harrison 1991; Biolsi and Zimmerman 1997; Churchill and Orelus 2012). This, however, should not deny or discredit the utopian intentions and liberatory efforts also involved in peasant research, but instead raise (ethical) questions over knowledge production in general, the (colonial/industrial) positionality embedded within that research, how that information can be used and what actors will have access to research insights (see Dunlap 2018e, 2019: 9–15). Peasant studies, like the Vietnam War, eventually died down, but would continue steadily alongside the overt, covert, preparatory and diplomatic interventions of the global superpowers.

Meanwhile, peasant studies had other recursive effects as well. Debates over 'modes of production'—and the question of whether rural social structures are 'feudal' or 'capitalist'—had 'real-world consequences' (Edelman and Wolford 2017: 6) as theoretical positions gave birth to strategic and tactical choices as these were incorporated by Communist movements in different parts of the world. In India, for example, such recursive effects are still present to this day as modes of production debates reverberate throughout Communist politics (Lerche et al. 2013). In a rare example of agrarian research on the contemporary Maoist (Naxalite) guerrilla movement operating among Indigenous populations in hilly and forested parts of India, Alpa Shah (2013) shows how the Communist guerrillas' theoretical positions, manifested both in theoretical manifestos (CPI Maoist 2004) and praxis, constrain their ability to analyze agrarian relations on the ground (see also Jakobsen 2016, 2017). Agrarian studies, we can say, is charting the developmental aspirations, conflicts and byproducts of what Perlman (2010 [1983]) called 'worms', 'Octopuses' and Leviathanic 'beasts' that progress in the direction of forming Worldeater(s).

Political Ecology

Peasant research would inspire and continue to develop alongside the emerging field of political ecology in the 1980s. While 'political ecology' was first used by Frank Thone (1935), the term did not develop until the second half of the twentieth century. With academic roots in peasant studies, cultural ecology and hazard studies, the term 'political ecology' was then employed in an article title by Eric Wolf (1972), but not in text. Shortly after, Enzensberger (1974) used the term to refer to the environmental movement of the 1960–70s, 'which he saw as fundamentally rooted in capitalist techno-science and therefore incapable of addressing the structural causes of environmental crises' (Perreault et al. 2015: 4). The field of political ecology was nourished by decolonization movements, the Vietnam War and subsequent CIA proxy wars overseas, which unfolded alongside inner city uprisings, student, anti-war, deep ecology and the anti-nuclear moments in the West. Political ecology would critique the simplicity of the environmental movement's Nature versus Civilization dichotomy, the popular obsessions with overpopulation,[2] resource scarcity[3] and, most importantly, the study of ecology's claim to scientific objectivity that masked its class, race, gendered and statist political interests with a so-called apolitical ecology (Robbins 2012 [2004]; Perreault et al. 2015; Dunlap and Brock 2019). The Worldeater, in many ways, echoes the nature vs civilization dichotomy, yet dispelling many of these myths, it recognizes humans as nature and, instead, focuses on the negative byproducts—'the shit'—of particular human actions as the product of civilization. The Worldeater framing thus stems in part from similar foundational concerns, while taking some of political ecology's oppositional positions a step further.

Political ecology theoretically drew (and still draws) on a wide range of disciplines, especially disciplines methodologically oriented toward field-based research methods (Perreault et al. 2015: 5). The seminal works in the Anglophone tradition of political ecology are traced to the 1980s with Watts (1983a, 1983b), Blaikie (1985) and Blaikie and Brookfield (1987), yet like peasant studies, political ecology was both influenced by the US Department of Defense (DoD) and Marxism. Political ecology was not separate from new found interests in peasants and revolutionary move-

[2] See *The Limits to Growth Report* (1972).

[3] See Garret Hardin's (1968) 'Tragedy of the Commons'.

ments across the world. '[S]tate-funded research initiatives, such as the US government's Fulbright programs, both encouraged and facilitated international research', write Perreault *and colleagues*, (2015: 5), as 'the US Department of Education's area studies programs (known as "National Resource Centers," or "Title VI" programs) were initially established by Title VI of the National Defense Education Act of 1958, and were thus closely aligned with US foreign policy objectives'. This, however, does not change the radical roots and politically conscious orientation of political ecology, which remained influenced by various types of Marxism(s). Perreault et al. (2015: 6) point out the influence of Marxist geography with David Harvey's (1974) critique of neo-Malthusianism in the liberal environmental movement and Neil Smith's (2010 [1983]) *Uneven Development* that were 'of enormous (though often unacknowledged) influence in political ecology.' This also included Harvey's (2018 [1982]) *Limits to Capital* that recognized that capitalist production of natural resources cannot be separated from the relationships of social production. These shared Marxian roots, however, tend to pave over the embedded anti-authoritarian and anarchistic tradition in political ecology, which, likewise, increasingly shines through critical agrarian studies.

Critical Agrarian Studies and Political Ecology: Difference and Convergence

Critical agrarian studies in its present form emerges out of the 2006–2008 'convergence of crises'—finance, food, climate, energy (McMichael 2012)—in the global capitalist economy and, relatedly, new re-valuations of natural resources with the rise of the green economy. Continuing the tradition of peasant studies, critical agrarian studies resituate the 'agrarian question' within 'dynamic and recurrent manifestations of multifaceted and contradictorily changing patterns of social and economic relations that continually and complexly reconfigure rural labour regimes' (Akram-Lodhi and Kay 2010b: 179–80; Levien et al. 2018). In essence, critical agrarian studies emerges to re-examine the impacts of neoliberal governance and economy on agrarian systems, dissecting new forms of technological integration, organizational forms, enclosure and commodification processes. This has brought critical agrarian studies to lead the discussion on land grabbing (Borras et al. 2011, 2012; White et al. 2012; Aguilar-Støen 2016; Levien 2018), which raised questions about land control (Peluso and Lund 2011) and exclusion (Hall et al. 2011). This includes the methodologies of land acquisition accounting (Scoones et al. 2013), state involvement in land grabs (Wolford et al. 2013; Lund 2016; Aguilar-Støen 2016), the difference between land grabs and land deals (Borras and Franco 2013; Schoenberger et al. 2017) and more dispersed or diffuse processes of dispossession (Vijayabaskar and Menon 2018; Vorbrugg 2019; Jakobsen and Nielsen in press). Moreover is analyzing the variegated 'political reactions "from below"' responding to land grabs (Borras and Franco 2013; Hall et al. 2015). The recent intensification of land grabbing has deeply influenced new agricultural arrangements (White et al. 2012; Fairbairn et al. 2014), methods of contracting (Li 2011; Lund 2018) and new 'green' or 'sustainable' resource valuations giving rise to biofuels (Borras et al. 2010), 'flex crops' (Oliveira and Hecht 2016; Borras et al. 2016) and 'flex trees' (Kröger 2016), eco-tourism and conservation in general (Kelly 2011; Ojeda 2012; Holmes 2014; Montenegro de Wit 2017; Fletcher et al. 2018), which is increasingly tied up with the emergence of 'new "climate change commodities"' (Dunlap and Fairhead 2014: 938; Hunsberger et al. 2017). This has led to the acknowledgment of not only land grabbing, but also 'green grabbing' (Fairhead et al. 2012; Corson et al. 2013; Franco and Borras 2019), 'value grabbing' (Andreucci et al. 2017) and ocean/blue grabbing (Benjaminsen and Bryceson 2012; Barbesgaard 2018), all of which 'involves transfers of the control of land and/or natural resources to powerful actors by various means using an environmental ethic or rationale' to justify land use, transformation, and often, destruction (Dunlap 2017: 17). The emphasis on land, especially in the matters of green grabbing, increasingly blurs an already thin line between critical agrarian studies and political ecology.

Political ecology, on the other hand, retains a general open-ended approach to investigating landscapes, natural resource management, socio-ecological relationships and environments more broadly. While proponents of critical agrarian studies recently acknowledged 'the need to bring urban and rural optics together, going beyond rural–urban linkages to see "nature in the city" and urbanized planning logics in the countryside' (Edelman and Wolford 2017: 15; see also Chari 2004; Cowan 2018), this line of research has existed in urban political ecology to various degrees (Swyngedouw 1996; Heynen et al. 2006), but especially so with rural gentrification studies that examine influxes of high-income (or higher-income) populations into rural areas (Phillips 1993; Smith 2011; Dunlap 2017). Rural gentrification is linked to new resource valuations that remain intimately tied to consumer trends, tourism, megaproject development, the green economy in general and, we would add, extractivist infrastructures across the rural world as Worldeater(s) body forms.

Natural resource management, like critical agrarian studies, entails researching dynamic and 'changing patterns of social and economic relations', which extends to investigating different epistemologies, ontological relationships with the land and interactions with more-than-human natures (see Blaser 2013; Escobar 2006, 2018; Sullivan 2017). Substantial research has been dedicated to natural resource governance, thus focusing on various agrarian (Agrawal and Gibson 1999; Agrawal 2005), but also conservation regimes and practices (Brockington 2002; West et al. 2006; Agrawal and Redford 2009), which has spawned the study of 'neoliberal natures' (Brockington and Duffy 2010; Sullivan 2006, 2010; Igoe 2010; Büscher et al. 2012; Duffy 2015). Neoliberal natures is inspired by Marxian and Foucauldian approaches to analyzing the infusion of neoliberal governmental practices and market inflected systems into ecological and land management schemes. There remains a significant and blurring overlap between critical agrarian studies and political ecology on this matter, yet political ecology has taken greater theoretical depth and focus on issues of environmental governance: 'eco-governance' (Ulloa 2013 [2005]), 'environmentality' (Agrawal 2005; Fletcher 2010, 2017) and 'greenmentality' (Cavanagh and Benjaminsen 2017). Aside from urban research, the greatest development of political ecology has been in the area of natural resource extraction, specifically hydrocarbon (Watts 1983b, 2004; McNeish 2015; Huber 2009; Bebbington and Bury 2013) and mineral extraction (Gudynas 2009; Bebbington et al. 2008; Bebbington 2012; Lang and Mokrani 2013 [2011]) as well as corresponding infrastructure such as pipelines and roads (see Bebbington and Bury 2013; Hindery 2013; McNeish 2015; Uribe 2018; Enns 2019). Significantly, political ecology has increasingly started examining—alongside critical agrarian studies (see Peluso 2017)—the confluence of conventional and green natural resource extraction (Sullivan 2013b; Büscher and Davidov 2013; Bury and Norris 2013; Duffy 2015; Dunlap 2017, 2018a), an issue taken up in greater depth in Chap. 5.

The sustained interest in governmentality and other Foucauldian influences in political ecology have been nourished by its 'sensitivity to representation, both as a set of discourses and as a field of practice' (Watts et al. 2011: 31). While the confluence of sensitivity to representation and abiding work with political economy has been definitional to political ecology, CAS tends more toward the latter stream of work. The strong influence of governmentality studies in political ecology points, moreover, to its abiding interest in conceptualizing the state, likewise explored discursively as well as materially (see Loftus 2018). In terms of extraction, this is registered in proliferating work on what Bridge (2014) calls the 'resource-state nexus'. In CAS, as Vergara-Camus and Kay (2017: 242) write, the 'recent literature on agrarian transitions and questions, land grabbing, green grabbing, and the rise of agribusiness have all highlighted the central role that the state plays', yet this has not led to commensurate attention being paid to conceptualizing the state as such. The Worldeater framing offered in Chap. 2, we could say, points to the need for more sustained work that goes beyond the 'resource-state nexus' in drawing on anarchistic thought for conceiving the state in terms of the colony model to explore the state as internal to the viral spread of the Worldeater's material grid.

This leads us to another noticeable difference from critical agrarian studies, namely a distinct anarchist influence on political ecology. Critical agrarian studies is by no means exempt from this influence, as James Scott (2009, 2012, 2017) makes clear, yet the Marxian foundations, while present in political ecology are de-centered. This de-centering takes three noticeable routes: anarchist theory, direct action and post-structuralism. While both CAS and political ecology are deeply rooted within social movements, the political roots are different (yet are increasingly leading to the same place). Peasant studies was imprinted with Marxism, peasant organizing and rural rebellion, which in a highly developed state would lead to Marxist-Leninist and Maoist armed formations. Critical agrarian studies, while retaining these roots to some degree, takes on an increasingly more Gramscian inspired approach (Castellanos-Navarrete and Jansen 2017; Li 2014; Jakobsen 2018a, b), an inspiration also found in political ecology (Mann 2009; Loftus 2013). More importantly, however, were rural Marxian inspired social movements, such as Via Campasina and the Landless Workers Movement[4] in Brazil, which take on more autonomous and, subsequently, anarchistic leanings such as with the Zapatisas (Vergara-Camus 2009, 2014; EZLN 2016) and other older Indigenous groups fighting for self-determination (Scott 2009; Dunlap 2018b, 2019c). While there have always been 'fractured' debates between authoritarian Marxists themselves (see Edelman and Wolford 2017), compounded by disagreements with anti-authoritarian Marxists (most notable between Lenin and Luxemburg), there still remains severe ontological dif-

[4] Trabalhadores Rurais Sem Terra.

ference between Marxists—even autonomous Marxists—and anarchists as to how they relate to power (see Springer 2016a, 2017; MTC 2018 [2016]; Dunlap 2018e), notably seizing the state apparatus or destroying it and dispersing power.

Robbins (2012: 25) explains that the anarchist Peter Kropotkin 'was an early political ecologist' with five discernable influences on the field of political ecology: (1) recognizing '*production* (farming, fishing, herding) as a key social-environmental process;' (2) 'a rigorous *archival and field-based* empirical approach' to processes of socio-ecological change; (3) 'an explicit concern for marginalized and disenfranchised communities;'(4) 'interest in the position and power of traditional environmental knowledge' in relation to social and technological change; and (5) 'starting from the landscape' when inquiring into socio-environmental issues. Taking Kropotkin's insights even further, Clark and Martin (2013 [1830]) refuted the human supremacy (or speciesism) implicit in industrial/enlightened regimes (see also Springer 2019; Dunlap and Brock 2019). While having greater influence in the francophone world, Reclus had an immense influence on eco-anarchism (see Clark and Martin (2013 [1830])). Following Reclus, Bookchin (2006: 19) developed the study of social ecology that 'is based on the conviction that nearly all of our present ecological problems originate in deep-seated social problems', where hierarchy is located at the roots of ecological disaster. Animating Bookchin's (1982) point, Kirkpatrick Sale (1991 [1985]: 122) explains:

Societies that dominate nature also dominate people. Where there is the idea that a massive dam should be built to control a river's flow, there is the idea that people should be enslaved to build it; where there is the belief that a giant metropole may serve itself by despoiling the surrounding countryside and devouring its raw materials, there are castes and hierarchies to ensure that this is accomplished.

The anarchist critique of domination and hierarchy has been hugely influential in environmental social movements and political ecology itself.

There exists, however, three lesser acknowledged, yet influential back-door anarchist influences on political ecology: 'direct action', Ivan Illich and Foucauldian post-structuralism. Missing in Robbins' list above is the principle of 'direct action' that is, as Heynen and van Sant (2015: 173) explain, a type of politics 'thinking outside of the state', rejecting hierarchical relations and engaging in unmediated political action, which has 'in fact been central to significant historical and social change for more than a century and a half'. Anarchist politics and intervention have been fundamentally influential on environmental politics, social struggle and the study of political ecology itself. This pillar in political ecology, however, has faded or remains in the shadows. Missing the heritage of direct action in political ecology, Springer (2019: 10) contends: 'Yet when it comes to anarchism and the use of direct action tactics… political ecology has been slow to respond'. It might be more accurate to say that political ecologists let this principle die as direct action in environmental and political activism began to take second fiddle to publishing and academic development as middleclass life (and problems) and (neoliberal) university restructuring intends. Needless to say, this is a trend that should be reversed, especially as academic and political conditions are worsening everywhere—some worse than others—across the world. Direct action, as we will explain in Chap. 6, can also be key to devising modes of resistance against the techno-capitalist Worldeater aiming at total liberation.

Ivan Illich would agree on this reversal or imbalance between academics and direct action. A silent partner in political ecology, Illich was an ardent critic of development (1970 [1969]), statist institutions (1970, 1973a, 1978a), the medical industry (1972), technology (1973b) and industrial systems for dispensing 'disabling professionals' (1978b), 'paralyzing affluence' and dependency on politico-economic systems (1978b: 11). 'Beyond a certain threshold', Illich (1978a: 10) writes, 'the multiplication of commodities induces impotence, the incapacity to grow food, to sing, or to build'. Sharing Taussig's (1980) concerns in the introduction, Illich (1978a) radically questioned the process of industrialization—both capitalist and communist—and dedicated a great deal of time to dissecting the dark side of developmental 'progress'. Defying identitarian labels, Illich's intellectual hostilities resonate with the iconoclastic individual anarchist Max Stirner (2017 [1845]), yet in his socio-institutional criticism, Illich speaks to the anarcho-communist tradition of Michael Bakunin and Kropotkin (see Watt 1981). Illich remained particularly influential and a strong influence on political ecology through well recognized authors: Arturo Escobar (2012 [1995]), Gustavo Esteva (2014) and Esteva and Madhu Suri Prakash (2014 [1997]), Eduardo Gudynas *and associates* (see Lang and Mokrani 2013 [2011]) as well as political ecologists associated with the Degrowth movement (see D'Alisa et al. 2014; Kallis 2018), who continue to develop Illich's theoretical insights through post-structuralism and Indigenous social movements.

Thirdly, there are various lines of anarchist influence on post-structuralism (Newman 2001; Rousselle and Evren 2011). As is well known, Foucault's post-structuralism has deeply influenced a plethora of political ecologists (see Peet et al. 2011). While Foucault (1991 [1981], 2003 [1997]: 10–3, 100) criticized Marx and said 'you will always find a racist component' in anarcho-communism and socialism (2003: 261), his project demonstrates a distinct affinity with the individualist anarchist tradition (Newman 2003; Rousselle and Evren 2011). The central themes of post-structuralism, such as 'the subject as flux and becoming, the instability of all identities, the critique of humanism and the rejection of the metaphysics of presence' all 'find their original and most forceful articulation in Stirner, even though his proximity is never really acknowledged', explains Newman (2011: 3). This even extends to Foucault's unspoken development of Stirner's (2017 [1845]: 204) ideas of power, 'prison society', fixation with the 'self' and Stirnarian 'intercourse' or 'sex'.[5] In short, the abiding association with post-structuralism and Foucauldian analytics embedded in political ecology to this day alerts us to the de-centering of Marxian categories in contradistinction to most work in critical agrarian studies.

CAS and Political Ecology: Recent Convergence and Direction

The recent emphasis on political reactions 'from below' in CAS, as mentioned earlier, has striking similarities with the political ecology of subsoil. The political reactions 'from below' recognizes that landless workers, elites, smallholders, Indigenous peoples, civil servants and so on, will (re) act in variegated ways when faced with incoming land deals and development projects. Borras *and colleagues* (Borras et al. 2012; Borras and Franco 2013; Hall et al. 2015) outline common trends associated with 'resistance, acquiescence or incorporation' and their intersections and divergences. While it is common for these three to operate simultaneously, there are roughly three types of conflicts.[6] The first, *people* versus *corporations*, pits

[5] With this in mind, we might venture to say, Foucault's (1998 [1978], 2007 [1976]: 162) interest in 'sex' in general, but as 'the hinge between anatomo-politics and bio-politics, it is at the intersection of disciplines and regulations, and it is in this function that it has become, at the end of the nineteenth century, a political drama of first importance for making society a machine of production' is building from and developing Stirner's (2017: 197) 'My Intercourse' chapter in his seminal work.

[6] Borras and colleagues both emphasize "poor people" within these conflicts, yet "People" can widen the possibilities of conflict dynamics, even if "poor people" are disproportionately negatively affected by land grabbing.

people against large companies seeking to control land for a desired business venture—agriculture, timber, mineral extraction, energy infrastructure—and is associated with various forms and intensities of land dispossession and/or the overall exploitation of both human and non-human natures. The second type, *people* versus *the state*, attests to the centrality of the state in facilitating, if not managing, land grabs (see Wolford et al. 2013; Lund 2016). The state is tasked with both inviting various consortiums and corporations to invest in large-scale land deals, while also maintaining popular legitimacy or, said differently, avoiding insurrection. The third, *people* versus *people*, relates to the splitting of various (class) interests[7] and (socio-cultural) value systems around different extractive projects. Development projects can embody the hopes, nightmares and indifferences of impacted populations, and hence 'resistance, acquiescence or incorporation'.

The political reactions 'from below' can be critiqued for not developing the multiplicity of political positions of resistance movements (see Nielsen 2018; Loadenthal 2017; Springer et al. 2019), instead prioritizing a mainstream or normative social movement perspective. Furthermore, this approach neglects the notion of political reactions 'from above' taken up by Verweijen and Dunlap (forthcoming) that investigates how conflict is managed and socially engineered by various state, corporate and elite actors. Nevertheless, the political reactions 'from below' has revealed the complications around land grabbing, significantly strengthening the conversation by detailing the various conflicting and contradictory actions taken by people when confronting development initiatives.

Meanwhile, there has been the emergence of the political ecology of the subsoil. Concerned with subsoil natural resource extraction, as opposed to CAS soil level, it nonetheless bares important similarities. The relative under-acknowledgment of these similarities, we would suggest, relates to the tendency to compartmentalize resources in separate 'sectors', a point that is acknowledged by Ian G. Baird and Keith Barney's (2017: 769) who point out that 'there has been a tendency to view terrestrial and aquatic ecosystems as areas requiring distinct research and expertise, and as a result the impacts of dams and plantation projects have tended to be assessed separately'[8]. The lack of attention to such similari-

[7] We would add race and gendered differences.

[8] As we noted in the Introduction, our lack of focus on aquatic resources is a limitation to this present exposition of total extractivism, to be fruitfully complemented by future work.

ties, moreover, can be seen as pointing more fundamentally to a lacking awareness of the 'nexus' to be explored in Chap. 5—that is, an unwillingness or reluctance to probe into the formation of techno-capitalist progress in the form of what we call the Worldeater. Bebbington and Bury (2013: 9–12) outline five foundations and practices for the political ecology of the subsoil. First, it advocates an increased interest in the natural sciences associated with subsoil extraction—biogeography, geology, hydrology, soil science, and so on—to better understand measurements (and their shortcomings), which will increase the effectiveness of political ecologists in understanding the socio-ecological impacts of extraction on populations. Second, recognizing the depth of the land transformation, or destruction, caused by subsoil extraction, the political ecology of the subsoil emphasizes the importance of learning how different populations respond, at different scales, to resource extraction. Third, it seeks to connect 'global production networks/value chains approaches with those that focus on regional and territorial dynamics' of natural resource governance and extraction (see Bridge 2008). Fourth, it emphasizes greater research into the centrality of the state in the political ecologies of extraction as the state is a powerful actor negotiating and determining patterns of natural resource access and control. Fifth, it pushes for more research into the enclosure process around subsoil extraction as the 'relationships between enclosure, commodification, and struggle therefore becomes central to understanding processes of landscape transformations in areas affected by extraction'.

While it might be argued this academic specificity does the opposite of what is intended with the Worldeater framing, these five points have clear overlaps with critical agrarian studies. The increased emphasis to engage with the natural sciences to improve measurements is similar to debates around methodological concerns around the reliability of data measuring land grabs (see Scoones et al. 2013). Following this, examining how populations respond to different extractive interventions aligns closely with the political reactions 'from below' research (rooted in pacifying peasant rebellions). Next, global production networks/value chains take on increasing importance especially with the rise of the green economy and renewable energy, which are increasingly implicated in hydrocarbon and mineral extraction (see Dunlap 2018c; Dunlap and Brock 2019), as discussed below. The fourth issue—the state—is already addressed by critical agrarian studies at some length (see Wolford 2013; Lund 2016) and political ecology (Loftus 2018), but could benefit from expanding its view (see

Newman 2001; Sharma and Gupta 2006; Springer 2016b). Finally, the fifth political ecology of the subsoil issue regarding the 'relationships between enclosure, commodification, and struggle' is, as we have seen, no less than the preeminent focus of critical agrarian studies. All of this is to suggest that emerging convergences on the manifold processes and technologies of extraction contributes to preparing political ecology and critical agrarian studies to delving further into the imperative of total extractivism and descend into the realm of the Worldeater(s).

This direction and overlap remains academically interesting, yet we would stress the imperative of stopping the Worldeater—techno-capitalist progress. We might ask skeptically, what will this research change in struggles to defend land and territory? How will it change socio-ecological crisis at this crucial moment in the history of the planet? Maybe these detailed studies will assist in court cases working to stop land grabs—which is very important and consequential, leave no doubt—but we would plead that research is organized to make legible the operations of the Worldeater or help position struggles aiming at stopping it. Knowledge production, we would think, is about improving the situation of the planet and its inhabitants—and while the promise of progress, modernization and development has claimed as much, in terms of socio-ecological harmony, it has failed. What it has done instead was opt for blind economic, statist and technological consolidation and development, or in a few words: growing the operations of the Worldeater. Yes, there have been medicines and great devices produced, but at what cost? The Worldeater epoch, we contend, demands us to get serious about developing knowledge to stop the growth trajectory, stop the imperative of total extractivism that is at the heart of capitalism's consumption of vitality. While we explore some ways of resisting the Worldeater in Chap. 6, we want to end this disciplinary overview offered here with a heartfelt invocation of the need—the utterly pressing demand—of developing ecological harmony. Taking a lesson from the genealogy presented in this chapter, ecological harmony must be genuine, not serving as cover for revived accumulation, neoliberal governance and counterinsurgency operations.

Conclusion

In sum, while there is divergence in disciplinary focus, political movements and theory between political ecology and critical agrarian studies, there also remains deep affinity and overlap in these same areas. The two camps are coming into increasing confluence, not only in research focus, but there are numerous prominent scholars (e.g. Joan Martinez-Alier, Nancy Peluso, James Fairhead, Tor Benjaminsen, Tania Murray Li) at the center of both fields. This increasing confluence invites reflections on how real the distinction between the two fields really is; as with intellectual division of labor generally, one may question its fruitfulness in grasping the catastrophic socio-ecological reality of total extractivism. To a certain extent, the very division may serve, primarily, the never-ending hunger of academic publishing and neoliberal university systems built around suffocating metrics. Linking the disciplinary confluence to the provocative thesis laid out in this book, moreover, we might suggest that for many of these scholars, their interest lies in understanding the shifts, changes and progression of the Worldeater, even if many political ecologists and critical agrarian researchers would not share this level of negativity toward industrial systems, nor use this type of abstract framing. Regardless, in line with the hostility of humans first encountering capitalist systems and with dissident anarchistic authors, we see value in naming the destruction process for what it is, and approaching it as such. This is not about advancing techno-capitalist progress as is the modus operandi of universities, but about stopping a monster—the Worldeater(s). It remains to be seen if academia can turn its back on this self-destructive progress—commodity fetishism, technological enchantment, statist dependency—and toward total life affirmation, position itself toward an insurrection against socio-ecological catastrophe. Chapter 4 will review and discuss two themes increasingly at the center of both critical agrarian studies and political ecology: conventional and green militarization—what we conceive as the 'claws and teeth' of the Worldeater(s). Chapter 5 will then delve into the discussion around the increasing collaboration between the conventional and green economy in order to chart the trajectory of total extractivism.

Bibliography

Agrawal, A. (2005). *Environmentality: Technologies of Government and the Making of Subjects*. Durham, NC: Duke University Press.

Agrawal, A., & Gibson, C. C. (1999). Enchantment and Disenchantment: The Role of Community in Natural Resource Conservation. *World Development, 27*, 629–649.

Agrawal, A., & Redford, K. (2009). Conservation and Displacement: An Overview. *Conservation and Society, 7*, 1–10.

Aguilar-Støen, M. (2016). Beyond Transnational Corporations, Food and Biofuels: The Role of Extractivism and Agribusiness in Land Grabbing in Central America. *Forum for Development Studies, 43*, 155–175.

Akram-Lodhi, A. H. (2018). What Is Critical Agrarian Studies? *Review of African Political Economy*. Retrieved from http://roape.net/2018/03/28/what-is-critical-agrarian-studies/.

Akram-Lodhi, A. H., & Kay, C. (2010a). Surveying the Agrarian Question (Part 1): Unearthing Foundations, Exploring Diversity. *Journal of Peasant Studies, 37*, 177–202.

Akram-Lodhi, A. H., & Kay, C. (2010b). Surveying the Agrarian Question (Part 2): Unearthing Foundations, Exploring Diversity. *Journal of Peasant Studies, 37*, 255–284.

Albertus, M., & Kaplan, O. (2013). Land Reform as a Counterinsurgency Policy: Evidence from Colombia. *Journal of Conflict Resolution, 57*, 198–231.

Anderson, K. B. (2010). *Marx at the Margins*. Chicago: University of Chicago Press.

Andreucci, D., García-Lamarca, M., Wedekind, J., et al. (2017). "Value Grabbing": A Political Ecology of Rent. *Capitalism Nature Socialism, 28*, 28–47.

Baird, I. G., & Barney, K. (2017). The Political Ecology of Cross-Sectoral Cumulative Impacts: Modern Landscapes, Large Hydropower Dams and Industrial Tree Plantations in Laos and Cambodia. *The Journal of Peasant Studies, 44*, 769–795.

Barbesgaard, M. (2018). Blue Growth: Savior or Ocean Grabbing? *The Journal of Peasant Studies, 45*(1), 130–149.

Bebbington, A. (2012). *Social Conflict, Economic Development and the Extractive Industry: Evidence from South America*. London: Routledge.

Bebbington, A., Bebbington, D. H., Bury, J., et al. (2008). Mining and Social Movements: Struggles Over Livelihood and Rural Territorial Development in the Andes. *World Development, 36*, 2888–2905.

Bebbington, A., & Bury, J. (2013). *Subterranean Struggles: New Dynamics of Mining, Oil, and Gas in Latin America*. Austin, TX: University of Texas Press.

Benjaminsen, T. A., & Bryceson, I. (2012). Conservation, Green/Blue Grabbing and Accumulation by Dispossession in Tanzania. *Journal of Peasant Studies, 39*(2), 335–355.

Bernstein, H. (2010). *Class Dynamics of Agrarian Change*. London: Kumarian Press.

Bernstein, H., Friedmann, H., van der Ploeg, J. D., et al. (2018). Fifty Years of Debate on Peasantries, 1966–2016. *The Journal of Peasant Studies, 45*, 689–714.

Biolsi, T., & Zimmerman, L. J. (1997). *Indians and Anthropologists: Vine Deloria, Jr., and the Critique of Anthropology*. Tucson, AZ: University of Arizona Press.

Blaikie, P. (1985). *The Political Economy of Soil Erosion in Developing Countries*. London: Routledge.

Blaikie, P., & Brookfield, H. (1987). *Land Degradation and Society*. London: Methuen.

Blaser, M. (2013). Notes Toward a Political Ontology of 'Environmental' Conflicts. In L. Green (Ed.), *Contested Ecologies: Dialogues in the South on Nature and Knowledge* (pp. 13–27). Cape Town: HSRC Press.

Bookchin, M. (1982). *The Ecology of Freedom: The Emergence and Dissolution of Hierarchy*. Palo Alto, CA: Cheshire Books.

Bookchin, M. (2006). *Social Ecology and Communalism*. Oakland, CA: AK Press.

Borras, S. M., & Franco, J. (2013). Global Land Grabbing and Political Reactions 'From Below'. *Third World Quarterly, 34*, 1723–1747.

Borras, S. M., Jr., Franco, J. C., Isakson, S. R., et al. (2016). The Rise of Flex Crops and Commodities: Implications for Research. *The Journal of Peasant Studies, 43*(1), 93–115.

Borras, S. M., Jr., Hall, R., Scoones, I., et al. (2011). Towards a Better Understanding of Global Land Grabbing: An Editorial Introduction. *The Journal of Peasant Studies, 38*, 209–216.

Borras, S. M., Kay, C., Gómez, S., et al. (2012). Land Grabbing and Global Capitalist Accumulation: Key Features in Latin America. *Canadian Journal of Development Studies/Revue canadienne d'e'tudes du de'veloppement, 33*, 402–416.

Borras, S. M., McMichael, P., & Scoones, I. (2010). The Politics of Biofuels, Land and Agrarian Change: Editors' Introduction. *Journal of Peasant Studies, 37*, 575–592.

Bridge, G. (2008). Global Production Networks and the Extractive Sector: Governing Resource-Based Development. *Journal of Economic Geography, 8*, 389–419.

Bridge, G. (2014). Resource Geographies II: The Resource-State Nexus. *Progress in Human Geography, 38*(1), 118–130.

Brockington, D. (2002). *Fortress Conservation*. Bloomington, IN: Indian University Press.

Brockington, D., & Duffy, R. (2010). Antipode Special Issue on Capitalism and Conservation. *Antipode, 42*, 469–484.

Bury, J., & Norris, T. (2013). Rocks, Rangers, and Resistance: Mining and Conservation Frontiers in the Cordillera Huayhuash, Peru. In A. Bebbington & J. Bury (Eds.), *Subterranean Struggles: New Dynamics of Mining, Oil, and Gas in Latin America*. Austin, TX: University of Texas Press.

Büscher, B., & Davidov, V. (2013). *The Ecotourism-Extraction Nexus: Political Economies and Rural Realities of (Un)Comfortable Bedfellows*. London: Routledge.

Büscher, B., Sullivan, S., Neves, K., et al. (2012). Towards a Synthesized Critique of Neoliberal Conservation. *Capitalism Nature Socialism, 23*, 4–30.

Castellanos-Navarrete, A., & Jansen, K. (2017). Why Do Smallholders Plant Biofuel Crops? The 'Politics of Consent' in Mexico. *Geoforum, 87*(Suppl. C), 15–27.

Cavanagh, C. J., & Benjaminsen, T. A. (2017). Political Ecology, Variegated Green Economies, and the Foreclosure of Alternative Sustainabilities. *Journal of Political Ecology, 24*, 200–216.
Chari, S. (2004). *Fraternal Capital: Peasant-Workers, Self-Made Men, and Globalization in Provincial India*. Stanford, CA: Stanford University Press.
Churchill, W., & Orelus, P. (2012). Confronting Western Colonialism, American Racism, and White Supremacy: Ward Churchill and Pierre Orelus in Dialogue. In P. W. Orelus (Ed.), *A Decolonizing Encounter: Ward Churchill and Antonia Darder in Dialogue* (pp. 56–112). New York: Peter Lang.
Clark, J., & Martin, C. (2013 [1830]). *Anarchy, Geography, Modernity: Selected Writings of Elisée Reclus*. Oakland, CA: PM Press.
Copeland, N. (2012). Greening the Counterinsurgency: The Deceptive Effects of Guatemala's Rural Development Plan of 1970. *Development and Change, 43*, 975–998.
Corson, C., MacDonald, K. I., & Neimark, B. (2013). Grabbing "Green": Markets, Environmental Governance and the Materialization of Natural Capital. *Human Geography, 6*, 1–15.
Cowan, T. (2018). The Urban Village, Agrarian Transformation, and Rentier Capitalism in Gurgaon, India. *Antipode, 50*, 1244–1266.
CPI (Communist Party of India) Maoist. (2004). *Party Programme*. Central Committee (P). Retrieved from http://www.bannedthought.net/India/CPI-Maoist-Docs/.
Cullather, N. (2006). "The Target Is the People": Representations of the Village in Modernization and U.S. National Security Doctrine. *Cultural Politics, 2*, 29–48.
D'Alisa, G., Demaria, F., & Kallis, G. (2014). *Degrowth: A Vocabulary for a New Era*. London: Routledge.
Duffy, R. (2015). Nature-Based Tourism and Neoliberalism: Concealing Contradictions. *Tourism Geographies, 17*, 529–543.
Dunlap, A. (2017). 'The Town Is Surrounded': From Climate Concerns to Life Under Wind Turbines in La Ventosa, Mexico. *Human Geography, 10*, 16–36.
Dunlap, A. (2018a). Insurrection for Land, Sea and Dignity: Resistance and Autonomy Against Wind Energy in Álvaro Obregón, Mexico. *Journal of Political Ecology, 25*, 120–143.
Dunlap, A. (2018b). "A Bureaucratic Trap": Free, Prior and Informed Consent (FPIC) and Wind Energy Development in Juchitán, Mexico. *Capitalism Nature Socialism, 29*, 88–108.
Dunlap, A. (2018c). *Reconsidering the Logistics of Autonomy: Ecological Autonomy, Self-Defense and the Polícia Comunitaria in Álvaro Obregón, Mexico*. Retrieved from https://www.tni.org/files/article-downloads/erpi_cp_8_dunlap.pdf.
Dunlap, A. (2018d). *End the "Green" Delusions: Industrial-Scale Renewable Energy Is Fossil Fuel+*. Retrieved from https://www.versobooks.com/blogs/3797-end-the-green-delusions-industrial-scale-renewable-energy-is-fossil-fuel.

Dunlap, A. (2018e). Book Review: The Anarchist Roots of Geography: Toward Spatial Emancipation by Simon Springer. *Human Geography, 11*, 62–64.

Dunlap, A. (2019). *Renewing Destruction: Wind Energy Development in Oaxaca, Mexico*. London: Rowman & Littlefield.

Dunlap, A. (2019c). Revisiting the Wind Energy Conflict in Gui'Xhi' Ro/Álvaro Obregón: Interview with an Indigenous Anarchist. *Journal of Political Ecology, 26*, 150–166.

Dunlap, A., & Brock, A. (2019). When the Wolf Guards the Sheep: Green Extractivism in Germany and Mexico. In S. Springer, M. Locret, J. Mateer, et al. (Eds.), *Anarchist Political Ecology* (Vol. 3). Oakland, CA: PM Press.

Dunlap, A., & Fairhead, J. (2014). The Militarisation and Marketisation of Nature: An Alternative Lens to 'Climate-Conflict'. *Geopolitics, 19*, 937–961.

Edelman, M., Oya, C., & SMB, J. (2013). Global Land Grabs: Historical Processes, Theoretical and Methodological Implications and Current Trajectories. *Third World Quarterly, 34*, 1517–1531.

Edelman, M., & Wolford, W. (2017). Introduction: Critical Agrarian Studies in Theory and Practice. *Antipode, 49*, 959–976.

Enns, C. (2019). Infrastructure Projects and Rural Politics in Northern Kenya: The Use of Divergent Expertise to Negotiate the Terms of Land Deals for Transport Infrastructure. *The Journal of Peasant Studies, 46*, 358–376.

Enzensberger, H. M. (1974). A Critique of Political Ecology. *New Left Review, 84*, 3–32.

Escobar, A. (2006). Difference and Conflict in the Struggle over Natural Resources: A Political Ecology Framework. *Development, 49*, 6–13.

Escobar, A. (2008). *Territories of Difference: Place, Movements, Life, Reads*. Durham: Duke University Press.

Escobar, A. (2012 [1995]). *Encountering Development: The Making and Unmaking of the Third World*. Princeton, NJ: Princeton University Press.

Escobar, A. (2018). *Designs for the Pluriverse: Radical Interdependence, Autonomy, and the Making of Worlds*. Durham, NC: Duke University Press.

Esteva, G. (2014). Commoning in the New Society. *Community Development Journal, 49*, 144–159.

Esteva, G., & Prakash, M. S. (2014). *Grassroots Postmodernism: Remaking the Soil of Cultures*. London: Zed Books Ltd.

EZLN. (2016). *Critical Thought in the Face of the Capitalist Hydra: I: Contributions by the Sixth Commission of the EZLN*. Brisbane: Paper Boat Press.

Fairbairn, M., Fox, J., Isakson, S. R., et al. (2014). Introduction: New Directions in Agrarian Political Economy. *Journal of Peasant Studies, 41*, 653–666.

Fairhead, J., Leach, M., & Scoones, I. (2012). Green Grabbing: A New Appropriation of Nature? *Journal of Peasant Studies, 39*, 237–261.

Fletcher, R. (2010). Neoliberal Environmentality: Towards a Poststructuralist Political Ecology of the Conservation Debate. *Conservation and Society, 8*, 171.

Fletcher, R. (2017). Environmentality Unbound: Multiple Governmentalities in Environmental Politics. *Geoforum, 85*, 311–315.

Fletcher, R., Dressler, W. H., Anderson, Z. R., et al. (2018). Natural Capital Must Be Defended: Green Growth as Neoliberal Biopolitics. *The Journal of Peasant Studies, 46*, 1–28.

Foucault, M. (1998 [1978]). *The Will to Knowledge: The History of Sexuality: 1.* London: Penguin Books.

Foucault, M. (2003 [1997]). *"Society Must Be Defended": Lectures at the College de France 1975–1976.* New York: Picador.

Foucault, M. (2007 [1976]). The Meshes of Power. In J. W. Crampton & S. Elden (Eds.), *Space, Knowledge and Power: Foucault and Geography.* Farnham: Ashgate Publishing Limited.

Foucault, M. (2007 [1978]). *Security, Territory, Population: Lectures at the College de France 1977–1978.* New York: Picador.

Franco, J. C., & Borras, S. M. (2019). Grey Areas in Green Grabbing: Subtle and Indirect Interconnections Between Climate Change Politics and Land Grabs and Their Implications for Research. *Land Use Policy, 84*, 192–199.

Gudynas, E. (2009). Diez tesis urgentes sobre el nuevo extractivismo. In J. Schuldt, A. Acosta, A. Barandiarán, et al. (Eds.), *Extractivismo, política y sociedad.* Quito: Centro Andino de Acción Popular (CAAP) and Centro Latinoamericano de Ecología Social (CLAES).

Hall, R., Edelman, M., Borras, S. M., et al. (2015). Resistance, Acquiescence or Incorporation? An Introduction to Land Grabbing and Political Reactions 'From Below'. *Journal of Peasant Studies, 42*, 467–488.

Hardin, G. (1968). The Tragedy of the Commons. *Science, 162*, 1234–1248.

Harrison, F. V. (1991). *Decolonizing Anthropology Moving Further Toward an Anthropology for Liberation.* Washington, DC: American Anthropological Association.

Harvey, D. (1974). Ideology and Population Theory. *International Journal of Health Services, 4*(3), 515–537.

Harvey, D. (2018 [1982]). *The Limits to Capital.* London: Verso.

Heynen, N., Kaika, M., & Swyngedouw, E. (Eds.). (2006). *The Nature of Cities: Urban Political Ecology and the Politics of Urban Metabolism.* New York: Routledge.

Heynen, N., & Robbins, P. (2005). The Neoliberalization of Nature: Governance, Privatization, Enclosure and Valuation. *Capitalism Nature Socialism, 16*, 5–8.

Heynen, N., & Sant, L. V. (2015). Political Ecologies of Activism and Direct Action Politics. In T. Perreault, G. Bridge, & J. McCarthy (Eds.), *The Routledge Handbook of Political Ecology.* London: Routledge.

Hindery, D. (2013). *From Enron to Evo: Pipeline Politics, Global Environmentalism, and Indigenous Rights in Bolivia.* Tucson, AZ: University of Arizona Press.

Holmes, G. (2014). What Is a Land Grab? Exploring Green Grabs, Conservation, and Private Protected Areas in Southern Chile. *Journal of Peasant Studies, 41*, 547–567.

Huber, M. T. (2009). Energizing Historical Materialism: Fossil Fuels, Space and the Capitalist Mode of Production. *Geoforum, 40*, 105–115.

Hunsberger, C., Corbera, E., Borras, S. M., Jr., et al. (2017). Climate Change Mitigation, Land Grabbing and Conflict: Towards a Landscape-Based and Collaborative Action Research Agenda. *Canadian Journal of Development Studies/Revue canadienne d'études du développement, 38*, 305–324.

Igoe, J. (2010). The Spectacle of Nature in the Global Economy of Appearances: Anthropological Engagements with the Spectacular Mediations of Transnational Conservation. *Critique of Anthropology, 30*, 375–397.

Illich, I. D. (1970 [1969]). *Planned Poverty: The End Result of Technical Assistance. Celebration of Awareness: A Call for Institutional Revolution*. London: Marion Boyars.

Illich, I. (1972). *Limits to Medicine: Medical Nemesis—The Expropriation of Health*. London: Marion Boyars.

Illich, I. (1973a). *Energy and Equity*. London: Calder & Boyars Ltd.

Illich, I. (1973b). *Tools for Conviviality*. London: Calder & Boyars.

Illich, I. (1978a). *The Right to Useful Unemployment and Its Professional Enemies*. London: Marion Boyars.

Illich, I. (1978b). *Towards a History of Needs*. New York: Pantheon books.

Jakobsen, J. (2016). Disappearing Landlords and the Unmaking of Revolution: Maoist Mobilization, the State and Agrarian Change in Northern Telangana. In K. B. Nielsen & A. G. Nilsen (Eds.), *Social Movements and the State in India: Deepening Democracy?* (pp. 239–267). London: Palgrave Macmillan.

Jakobsen, J. (2017). Inside and Outside the State: Imagining the Indian State Among Subalternists and Naxalites. In P. Basu (Ed.), *Naxalite Politics: Post-Structuralist, Postcolonial and Subaltern Perspectives* (pp. 283–299). Kolkata: Setu Prakashani.

Jakobsen, J. (2018a). Neoliberalising the Food Regime 'Amongst Its Others': The Right to Food and the State in India. *The Journal of Peasant Studies*, 1–21.

Jakobsen, J. (2018b). Towards a Gramscian Food Regime Analysis of India's Agrarian Crisis: Counter-Movements, Petrofarming and Cheap Nature. *Geoforum, 90*, 1–10.

Jakobsen, J., & Nielsen, K. B. (forthcoming). Compounding Aspirations: Grounding Hegemonic Processes in India's Rural Transformations. *Canadian Journal of Development Studies/Revue canadienne d'études du développement*.

Kallis, G. (2018). *Degrowth*. New York: Columbia University Press.

Kautsky, K. (1988 [1899]). *The Agrarian Question*, 2 Vols. London: Zwan Publications.

Kelly, A. (2011). Conservation Practice as Primitive Accumulation. *The Journal of Peasant Studies, 38*, 683–701.

Kröger, M. (2016). The Political Economy of 'Flex Trees': A Preliminary Analysis. *The Journal of Peasant Studies, 43*(4), 886–909.

Lang, M., & Mokrani, D. (2013 [2011]). *Beyond Development: Alternative Visions from Latin America*. Amsterdam: Transnational Institute.

Le Billon, P. (2001). The Political Ecology of War: Natural Resources and Armed Conflicts. *Political Geography, 20*, 561–584.

Le Billon, P. (2012). *Wars of Plunder: Conflicts, Profits and the Politics of Resources.* New York, Columbia University Press.

Lenin, V. I. (1964 [1899]). *The Development of Capitalism in Russia.* Moscow: Progress Publishers.

Lerche, J., Shah, A., & Harriss-White, B. (2013). Introduction: Agrarian Questions and Left Politics in India. *Journal of Agrarian Change, 13*, 337–350.

Levien, M., Watts, M., & Yan, H. (2018). Agrarian Marxism. *The Journal of Peasant Studies, 45*(5–6), 853–888.

Li, T. M. (2011). Centering Labor in the Land Grab Debate. *Journal of Peasant Studies, 38*, 281–298.

Li, T. M. (2014). *Land's End: Capitalist Relations on an Indigenous Frontier.* Durham, NC: Duke University Press.

Light, J. S. (2003). *From Warfare to Welfare: Defense Intellectuals and Urban Problems in Cold War America.* London: The John Hopkins University Press.

Loadenthal, M. (2017). *The Politics of the Attack: Communiqués and Insurrectionary Violence.* Manchester: Manchester University Press.

Loftus, A. (2013). Gramsci, Nature and the Philosophy of Praxis. In M. Ekers, G. Hart, S. Kipfer, & A. Loftus (Eds.), *Gramsci: Space, Nature, Politics* (pp. 178–197). Oxford: Wiley-Blackwell.

Loftus, A. (2018). Political Ecology II: Whither the State? *Progress in Human Geography*, 1–11.

Lund, C. (2016). Rule and Rupture: State Formation Through the Production of Property and Citizenship. *Development and Change, 47*, 1199–1228.

Lund, C. (2018). Predatory Peace. Dispossession at Aceh's Oil Palm Frontier. *The Journal of Peasant Studies, 45*, 431–452.

Mann, G. (2009). Should Political Ecology Be Marxist? A Case for Gramsci's Historical Materialism. *Geoforum, 40*(3), 335–344.

Martínez-Alier, J. (2002). *The Environmentalism of the Poor: A Study of Ecological Conflicts and Valuation.* Northampton: Edward Elgar.

Martínez-Alier, J., & O'Connor, M. (1995). Ecological and Economic Distribution Conflicts. In R. Costanza, O. Segura, & J. Martínez-Alier (Eds.), *Getting Down to Earth: Practical Applications of Ecological Economics* (pp. 153–183). Washington, DC: Island press.

Martinez-Alier, J., Temper, L., Del Bene, D., et al. (2016). Is there a Global Environmental Justice Movement? *The Journal of Peasant Studies, 43*, 731–755.

Martínez-Torres, M. E., & Rosset, P. M. (2010). La Vía Campesina: The Birth and Evolution of a Transnational Social Movement. *The Journal of Peasant Studies, 37*, 149–175.

Marx, K. (1982 [1867]). *Capital: A Critique of Political Economy Vol. I.* London: Penguin Books.

McKay, B. M. (2017). Agrarian Extractivism in Bolivia. *World Development, 97*, 199–211.

McMichael, P. (2012). The Land Grab and Corporate Food Regime Restructuring. *The Journal of Peasant Studies, 39*(3–4), 681–701.

McNeish, J.-A., Borchgrevink, A., & Logan, O. (2015). *Contested Powers: The Politics of Energy and Development in Latin America*. London: Zed Books Ltd.

Montenegro de Wit, M. (2017). Beating the Bounds: How Does 'Open Source' Become a Seed Commons? *The Journal of Peasant Studies, 46*, 1–36.

Moore, B. (1966). *Social Origins of Dictatorship and Democracy: Lord and Peasant in the Making of the Modern World*. London: Beacon Press.

MTC (Mauvaise Troupe Collective). (2018 [2016]). *The Zad and NoTAV: Territorial Struggles and the Making of a New Political Intelligence*. London: Verso Books.

Neimark, B. (2012). Green Grabbing at the 'Pharm' Gate: Rosy Periwinkle Production in Southern Madagascar. *Journal of Peasant Studies, 39*, 423–445.

Newman, S. (2001). War on the State: Stirner and Deleuze's Anarchism. *Anarchist Studies, 9*, 147–164.

Newman, S. (2003). Stirner and Foucault: Toward a Post-Kantian Freedom. *Postmodern Culture, 13*(2).

Newman, S. (2011). *Max Stirner*. New York: Palgrave.

Nielsen, K. B. (2018). *Land Dispossession and Everyday Politics in Rural Eastern India*. London and New York: Anthem Press.

Ojeda, D. (2012). Green Pretexts: Ecotourism, Neoliberal Conservation and Land Grabbing in Tayrona National Natural Park, Colombia. *The Journal of Peasant Studies, 39*, 357–375.

Oliveira, G., & Hecht, S. (2016). Sacred Groves, Sacrifice Zones and Soy Production: Globalization, Intensification and Neo-Nature in South America. *The Journal of Peasant Studies, 43*(2), 251–285.

Peet, R., Robbins, P., & Watts, M. (2011). *Global Political Ecology*. New York: Routledge.

Peluso, N. L. (2017). Plantations and Mines: Resource Frontiers and the Politics of the Smallholder Slot. *The Journal of Peasant Studies, 44*, 834–869.

Peluso, N. L., & Lund, C. (2011). New Frontiers of Land Control: Introduction. *The Journal of Peasant Studies, 38*, 667–681.

Peluso, N. L., & Watts, M. (2001). *Violent Environments*. Ithaca, NY: Cornell University Press.

Perlman, F. (2010 [1983]). *Against His-Story, Against Leviathan*. Detroit: Red and Black Press.

Perreault, T., Bridge, G., & McCarthy, J. (2015). *The Routledge Handbook of Political Ecology*. London: Routledge.

Phillips, M. (1993). Rural Gentrification and the Processes of Class Colonisation. *Journal of Rural Studies, 9*, 123–140.

Polanyi, K. (2001 [1944]). *The Great Transformation: The Political and Economic Origins of Our Time*. Boston: Beacon Press.

Price, D. H. (2011). *Weaponizing Anthropology: Social Science in Service of the Militarized State*. Oakland, CA: AK Press/Counterpunch Books.
Price, D. H. (2014). Counterinsurgency by Other Names: Complicating Humanitarian Applied Anthropology in Current, Former, and Future War Zones. *Human Organization, 73*, 95–105.
Rasmussen, M. B., & Lund, C. (2018). Reconfiguring Frontier Spaces: The Territorialization of Resource Control. *World Development, 101*, 388–399.
Ribot, J., & Peluso, N. (2003). A Theory of Access. *Rural Sociology, 68*, 153–181.
Robbins, P. (2012 [2004]). *Political Ecology: A Critical Introduction*. Malden, MA: John Wiley & Sons.
Rosset, P. M., & Altieri, M. A. (2017). *Agroecology: Science and Politics*. Rugby: Practical Action Publishing.
Rosset, P. M., Machín Sosa, B., Roque Jaime, A. M., et al. (2011). The Campesino-to-Campesino Agroecology Movement of ANAP in Cuba: Social Process Methodology in the Construction of Sustainable Peasant Agriculture and Food Sovereignty. *The Journal of Peasant Studies, 38*, 161–191.
Rousselle, D., & Evren, S. (2011). *Post-Anarchism: A Reader*. London: Pluto Press.
Sale, K. (1991 [1985]). *Dwellers in the Land: The Bioregional Vision*. Philadelphia: University of Georgia Press.
Salemink, O. (2003). Social Science Intervention: Moral Versus Political Economy and the Vietnam War. In P. Q. Van Ufford & A. K. Giri (Eds.), *A Moral Critique of Development: In Search of Global Responsibilities*. London: Routledge.
Scheidel, A., Temper, L., Demaria, F., et al. (2017). Ecological Distribution Conflicts as Forces for Sustainability: An Overview and Conceptual Framework. *Sustainability Science, 13*, 585–598.
Schoenberger, L., Hall, D., & Vandergeest, P. (2017). What Happened When the Land Grab Came to Southeast Asia? *Journal of Peasant Studies, 44*, 697–725.
Scoones, I., Hall, R., Borras, S. M., et al. (2013). The Politics of Evidence: Methodologies for Understanding the Global Land Rush. *Journal of Peasant Studies, 40*, 469–483.
Scott, J. C. (1977). *The Moral Economy of the Peasant: Rebellion and Subsistence in Southeast Asia*. New Haven: Yale University Press.
Scott, J. C. (2009). *The Art of Not Being Governed: An Anarchist History of Upland Southeast Asia*. New Haven, CT: Yale University Press.
Scott, J. C. (2012). *Two Cheers for Anarchism: Six Easy Pieces on Autonomy, Dignity, and Meaningful Work and Play*. Princeton, NJ: Princeton University Press.
Scott, J. C. (2017). *Against the Grain: A Deep History of the Earliest States*. New Haven, CT: Yale University Press.
Shah, A. (2013). The Agrarian Question in a Maoist Guerrilla Zone. *Journal of Agrarian Change, 13*, 424–450.
Sharma, A., & Gupta, A. (2006). *The Anthropology of the State: A Reader*. London: John Wiley & Sons.

Smith, N. (2010 [1983]). *Uneven Development: Nature, Capital, and the Production of Space*. London: Verso.

Smith, D. P. (2011). What Is Rural Gentrification? Exclusionary Migration, Population Change, and Revalorised Housing Market. *Interface: Planning Theory & Practice, 12*, 596–605.

Solovey, M. (2001). Project Camelot and the 1960s Epistemological Revolution: Rethinking the Politics-Patronage-Social Science Nexus. *Social Studies of Science, 31*, 171–206.

Springer, S. (2016a). *The Anarchist Roots of Geography: Toward Spatial Emancipation*. Minneapolis: University of Minnesota Press.

Springer, S. (2016b). *Learning Through the Soles of Our Feet: Unschooling, Anarchism, and the Geography of Childhood*. London: Rowman & Littlefield.

Springer, S. (2017). The Limits to Marx: David Harvey and the Condition of Postfraternity. *Dialogues in Human Geography, 7*(3), 280–294.

Springer, S., Locret, M., Mateer, J., et al. (2019). *Anarchist Political Ecology*. Oakland, CA: PM Press.

Stirner, M. (2017 [1845]). *The Unique and Its Property*. Berkeley, CA: Ardent Press.

Sullivan, S. (2006). Elephant in the Room? Problematising 'New' (Neoliberal) Biodiversity Conservation. *Forum for Development Studies, 33*, 105–135.

Sullivan, S. (2010). 'Ecosystem Service Commodities'—A New Imperial Ecology? Implications for Animist Immanent Ecologies, with Deleuze and Guattari. *New Formations: A Journal of Culture/Theory/Politics, 69*, 111–128.

Sullivan, S. (2013b). After the Green Rush? Biodiversity Offsets, Uranium Power and the 'Calculus of Casualties' in Greening Growth. *Human Geography, 6*, 80–101.

Sullivan, S. (2017). What's Ontology Got to Do with It? On Nature and Knowledge in a Political Ecology of the 'Green Economy'. *Journal of Political Ecology, 24*, 217–242.

Svarstad, H., Benjaminsen, T. A., & Overå, R. (2018). Power Theories in Political Ecology. *Journal of Political Ecology, 25*, 350–363.

Swyngedouw, E. (1996). The City as a Hybrid: On Nature, Society and Cyborg Urbanization. *Capitalism Nature Socialism, 7*(2), 65–80.

Taussig, M. T. (1980). *The Devil and Commodity Fetishism in South America*. Durham, NC: University of North Carolina Press.

Tetreault, D. V. (2014). Mexico: The Political Ecology of Mining. In H. Veltmeyer & J. F. Petras (Eds.), *The New Extractivism: A Post-Neoliberal Development Model or Imperialism of the Twenty-First Century?* London: Zed Books.

Thompson, E. P. (1991 [1963]). *The Making of the English Working Class*. London: Penguin Books.

Thone, F. (1935). Nature Rambling: We Fight for Grass. *The Science Newsletter, 27*, 14.

Ulloa, A. (2013 [2005]). *The Ecological Native: Indigenous Peoples' Movements and Eco-Governmentality in Columbia*. London: Routledge.

Uribe, S. (2018). Illegible Infrastructures: Road Building and the Making of State-Spaces in the Colombian Amazon. *Environment and Planning D: Society and Space*, 1–19.

Vergara-Camus, L. (2009). The MST and the EZLN Struggle for Land: New Forms of Peasant Rebellions. *Journal of Agrarian Change, 9*, 365–391.

Vergara-Camus, L. (2014). *Land and Freedom: The MST, the Zapatistas and Peasant Alternatives to Neoliberalism*. London: Zed Books.

Vergara-Camus, L., & Kay, C. (2017). Agribusiness, Peasants, Left-Wing Governments, and the State in Latin America: An Overview and Theoretical Reflections. *Journal of Agrarian Change, 17*, 239–257.

Verweijen, J., & Dunlap, A. (forthcoming). The Evolving Techniques of Social Engineering, Land Control and Managing Protest Against Extractivism: Introducing Political (Re)Actions 'From Above'. *Political Geography*.

Vijayabaskar, M., & Menon, A. (2018). Dispossession by Neglect: Agricultural Land Sales in Southern India. *Journal of Agrarian Change, 18*, 571–587.

Vorbrugg, A. (2019). Not About Land, Not Quite a Grab: Dispersed Dispossession in Rural Russia. *Antipode*. https://doi.org/10.1111/anti.12523.

Walker, R. A. (2004). *The Conquest of Bread: 150 Years of Agribusiness in California*. New York: The New Press.

Watt, A. (1981). Illich and Anarchism. *Educational Philosophy and Theory, 13*, 1–15.

Watts, M. J. (1983a). On the Poverty of Theory: Natural Hazards Research in Context. In K. Hewitt (Ed.), *Interpretations of Calamity: From the Viewpoint of Human Ecology* (pp. 231–262). Boston: Allen & Unwin.

Watts, M. J. (1983b). *Silent Violence: Food, Famine, and Peasantry in Northern Nigeria*. Berkeley, CA: University of California Press.

Watts, M. (2004). Resource Curse? Governmentality, Oil and Power in the Niger Delta, Nigeria. *Geopolitics, 9*(1), 50–80.

Watts, M., Robbins, P., & Peet, R. (2011). Global Nature. In R. Peet, P. Robbins, & M. Watts (Eds.), *Global Political Ecology*. London and New York: Routledge.

West, P., Igoe, J., & Brockington, D. (2006). Parks and Peoples: The Social Impact of Protected Areas. *Annual Review of Anthropology, 35*, 251–277.

White, B., Borras, S. M., Hall, R., et al. (2012). The New Enclosures: Critical Perspectives on Corporate Land Deals. *Journal of Peasant Studies, 39*, 619–647.

Williams, R. (1973). *The Country and the City*. New York: Oxford University Press.

Wolf, E. (1972). Ownership and Political Ecology. *Anthropological Quarterly, 45*, 201–205.

Wolf, E. R. (1999 [1969]). *Peasant Wars of the Twentieth Century*. Omaha: University of Oklahoma Press.

Wolford, W., Borras, S. M., Hall, R., et al. (2013). Governing Global Land Deals: The Role of the State in the Rush for Land. *Development and Change, 44*, 189–210.

Zoomers, A. (2010). Globalisation and the Foreignisation of Space: Seven Processes Driving the Current Global Land Grab. *Journal of Peasant Studies, 37*, 429–447.

CHAPTER 4

Claws & Teeth: The Militarization of Nature

Abstract In this chapter, we bring total extractivism into view of the militarization of nature. This involves an extensive review of research into the violent technologies of extraction, including studies of the requisite levels of land control and territorialization for sustaining and accelerating the present techno-capitalist trajectory. Through this review, we seek to draw out the political reactions 'from above' from governmental, corporate and elite actors as it relates to the conventional and so-called green military, police and extra-judicial forces, but also 'soft' pacification techniques. In doing so, this chapter presents the violent domination and manipulation—the 'claws and teeth' of the Worldeater—underpinning total extractivism.

Keywords Militarization • Violence • Counterinsurgency • Pacification • Political ecology

A. Dunlap, J. Jakobsen, *The Violent Technologies of Extraction*,
https://doi.org/10.1007/978-3-030-26852-7_4

States, militarization and, consequently, capitalist economies could not exist without natural resources. This chapter will delve into the politics of the claws and teeth of the Worldeater: coercive violence and social pacification. The highest priority of the Worldeater, or the imperative of techno-capitalist industrial progress, is thus acquiring, transforming and controlling natural resources, frequently deemed a 'strategic', 'critical' or a 'national security' interest (see Crosby and Monaghan 2018). Controlling human and natural resources, as James Scott (1998) showed us, begins with making them legible: revealing the location, quantity and type of timber, mineral and hydrocarbons. Meanwhile, human resources are organized through names, addresses and other forms of identification (birth certificates, social security numbers, finger prints, DNA), permitting access to civil amenities, but also making it easy for tax collectors, collection agencies, police and the military to find you (Scott 1998; Graham 2011; FM3-24 2014: 10–1). Disciplinary and biopolitical interventions follow legibility (see Foucault 1995, 1998, 2003), whereby human and non-human natures are disciplined, commodified and categorized—regimented into timber colonies and populations—by governmental systems to satisfy the needs of statist development and economic growth (see Marx 1982; Polanyi 2001; Thompson 1991 [1963]; Anthony 2001 [1977]; Federici 2004; Foucault 2007; Perelman 2007). While resistance and insurrection are enduring features against statist structures—disrupting, subverting and modifying governmental action—the tension toward internal stability persists, consolidating the colonial process within (see Dunlap 2018b; Crosby and Monaghan 2018), before exporting a crude version overseas to acquire new territory, resources and markets (see Galeano 1997 [1973]; Rodney 2009 [1972]). Said bluntly, industrial development requires various intensities of ecocide, genocide and slavery in both North and South of the Globe (see Wolfe 2006; Moses 2008; Moses and Stone 2013; Short 2016), which in essence forms the mythical Worldeater in question.

In this chapter, we bring total extractivism into view of the militarization of nature. Doing so, we take heed of the fact that the natural resources 'upon which industrial societies stand is constructed in large part through the use and threatened use of armed violence', leading Liam Downey, Eric Bonds and Katharine Clark (2010: 437) to conclude succinctly that 'armed violence underpins the current ecological crisis'. The approach we take here, however, is markedly different from the popular idea of 'the resource curse'. The resource curse, Anthony Bebbington (2012: 6) explains, is associated with

> the overvaluation of exchange rates that reduce the competitiveness of other sectors of the economy; an increasing narrowing of the national economy and hence vulnerability to price swings; the enclave characteristics of an extractive sector that generates few multiplier effects; the generation of vast rents that induce political behavior oriented towards capturing those rents rather than governing well; the growth of violent conflict driven by the desire to capture rents; and the emergence of states whose primary pacts are with extractive companies rather than their citizenries.

Linking countries' vulnerability to armed violence and conflict to the prevalence of natural resource wealth, the 'thesis' of the resource curse has been frequently popularized by scholars like Paul Collier and others, although it has been refuted by recent meta-analyses (O'Brochta 2019). While blatantly false in aggregation, the thesis moreover seems to portray armed conflict over resources in highly simplistic terms as one of the 'lock in' between 'rational actors' in the form of state and so-called rebels. Such views do not square well with advances in political ecology, such as that of Peluso and Watts's (2001: 5) notion of 'violent environments' where they 'see violence as a site-specific phenomenon rooted in local histories and social relations yet connected to larger processes of material transformation and power relations.' Neither do they square with recent attempts in critical agrarian studies toward seeing war and violence 'in agrarian perspective' involving the analytical foci we have outlined in Chap. 3 revolving around changing regimes of labor, land and class (Cramer and Richards 2011). Seeking to push the literature beyond, while complementing, the foregrounding of political reactions 'from below' expressed in such an 'agrarian turn', we proceed in this chapter to review work dealing with the violence embedded in land control and territorialization that shapes inquiries into the political ecology of counterinsurgency (see Dunlap 2019a). Through this review, we seek to draw out the political reactions 'from above' from governmental, corporate and elite actors. While Chap. 5 then goes on to map the organization of the system of total extractivism with its intricate webs of infrastructures, the current chapter supplies the requisite view of the violent domination and manipulation—the claws and teeth of the Worldeater—underpinning total extractivism. Sustaining our argument is a radically different approach to the question of violence from what the popularized notion of 'triggers' for conflict in environments/resources has to offer.

The Violence of Land Control and Territorialization

Viewing the historical trajectory of capitalism through the prism of total extractivism, we are not surprised to find that 'the term territory has an association with fear and violence' (Elden 2010: 806–7). Indeed, the work on land control and territorialization reveals the centrality of violence in both 'green' (intensive agriculture, forestry and conservation) and conventional (mineral and hydro carbon) forms of extraction. 'Violence in parks, resettlement areas, and plantations', write Nancy Peluso and Christian Lund (2011: 676), 'is connected to productions of new environmental subjects and to state-making itself.' This violence is largely conducted by governmental, corporate and, consequently, extra-judicial forces: paramilitary groups, criminal gangs and assassins.[1] While violence is subjective, without a single definition (Springer and Le Billon 2016; Gelderloos 2013), an immense amount of research has been conducted across peace studies, anthropology and geography to chart the different types of violence. Here, we offer eight different types: First, *direct violence*, from being hit by a nightstick to being shot by various types of munitions, which implies overt coercive action seeking to enforce (political) submission to formal and informal authorities, drawing on an array of police, military, extra-judicial forces (Stephen 2000; Bourgois 2001). Second, *structural violence* is related to institutional arrangements and planning that discriminate, injure or slowly kill populations—directly or indirectly—by withholding basic needs or placing people in toxic environments (Galtung 1969; Bourgois 2001). Structural violence is 'environmentally embedded violence' that is also a '*slow violence*' with 'delayed destruction that is dispersed across time and space, an attritional violence that is typically not viewed as violence at all' (Nixon 2011: 7, 2). Third, *symbolic violence* employs symbols to affirm authority, subjugate whole or segments of populations, akin to statues and flags symbolizing racism, sexism or, generally, a form of domination over a person or people (Bourdieu 2001; Bourgois 2001). Fourth, *epistemic violence* marginalizes, discredits or attempts to erase methods of knowledge generation, such as Indigenous or hermetic sciences and knowledges (Spivak 1988; Castro-Gomez 2002; Marker 2003). Fifth, *everyday violence* is the type that occurs on a personal level at home or on the street, related to domestic violence, harassment and assault (see Scheper-Hughes 1992; Bourgois 2001). Sixth, *infrastructural violence* relates the manipulative function of architecture, infrastructure and city planning as it is designed to invoke discomfort, prevent mobility and securitize space (Rodgers and O'Neill 2012; Li 2018; see Chap. 5). Seventh, *bureaucratic violence* is the discomfort, social suffering and death arising from, or supported by, bureaucratic systems employed by governments, corporations, legal ('justice') procedures and procedural arrangements in general (Eldridge and Reinke 2018). Eight, *sustainable violence* is the use of renewable or 'sustainable' technologies in the service of armed forces, to multiply their capacity to dispense repression (or 'security') or make (false) claims of ecological sustainability (Dunlap 2017b). These eight types of violence can overlap and emerge together in

[1] This is detailed in the context of Mexico in great depth by Dawn Paley (2014) and Guadalupe Correa-Cabrera (2017).

various ways, as has been the case over controlling land and natural resources.

Violence, however, is only spreading in quantity and intensity with the rise of the green economy, which is, as Chap. 3 indicated, one of the favored new terrains for the Worldeater to find sustenance and body formation. New resource valuations emanating from the green economy are creating new resource frontiers. Following the insights from critical agrarian studies, Fairhead et al. (2012), develop the term 'green grabbing', after journalist John Vidal (2008), to refer to land grabbing in the name of 'green', 'sustainable' or environmentally friendly projects. Such projects include biofuels (Borras et al. 2010; Baka 2013), eco-tourism (Ojeda 2012), biodiversity conservation (Ybarra 2012, 2017), forest conservation (Benjaminsen and Bryceson 2012; Holmes 2014), wind energy development (Siamanta 2019; Dunlap 2019b), solar power (Rignall 2016; Siamanta 2017), hydropower (Finley-Brook and Thomas 2011) and really anything that can be branded as 'green' and ecologically sustainable. New resource frontiers hinge upon waves of violence and extractive territorialization (Peluso and Lund 2011; Rasmussen and Lund 2018; Lund 2016). Complementing Foucault's (2003; see Dunlap 2014a) technologies of colonization, Mattias Rasmussen and Christian Lund (2018: 391–6) explain that territorialization includes the re-articulation of 'political authority, citizenship and property' relations; the construction of 'boundaries and maps'; the deployment of 'law and bureaucracy'; and the 'enforcement' of epistemological, discursive and political impositions through physical and symbolic violence.

The rise of the green economy and green grabbing has generated a substantial amount of interest in what we could call, as a category, green militarization. Since 2014, we have seen cross-cutting literature deploying the terms 'green militarization', 'green violence', 'green wars' and 'the political ecology of counterinsurgency', the latter acknowledging how counterinsurgency techniques have been 'greened' and put into advancing market-based conservation and green grabbing. Examining South Africa's Kruger National Park, Elizabeth Lunstrum (2014: 817) defines green militarization 'as the use of military and paramilitary (military-like) actors, techniques, technologies, and partnerships in the pursuit of conservation'. Displaying the complications of securitizing conservation (see also Massé and Lunstrum 2015), Lunstrum (2014: 829) emphasized how conservation values merged with ideas about national sovereignty to legitimize the use of military force, spawning a conservation-induced 'arms race' where 'commercial poachers become better armed', which is matched tit-for-tat by park rangers, leading to an 'intensifying cycle of militarization'. We can regard this as another method by which the Worldeater spreads through landscapes and initiates or furthers the control and accumulation of natural resources.

Continuing this line of inquiry, Bram Büscher and Maano Ramutsindela (2016) introduce the term 'green violence' in the context of South African trans boundary 'peace parks' (with Zimbabwe and Mozambique) to describe the intensifying acts of militarization and violence committed in the name of Rhino protection. There are three elements to green violence. First, material violence seeks to show the rippling material effects of militarization and warfare as a mode of biodiversity conservation, which, to Büscher and Ramutsindela, transcends the limitations of green militarization. Second, social violence, following Arthur Kleinman (2000), are the social and moral orders emanating from (trusted) governmental and social institutions that protect and attack various segments of populations. The social violence in the case of biodiversity conservation emerges, for Büscher and Ramutsindela (2016: 13), by 'both harming the public moral order and the (ab)use of social power in pursuit of conservation-related ideas and aspirations.' Third, discursive violence is the popular, often catchy, narrative that constructs a problem or an enemy and, in this case, seeks to commit various acts of violence against them. Green violence is related to previous socio-political contexts, refusal to address previous histories of political violence and acts out a colonial political economy. While 'warfare ecology' was coined to further research the ecological impact of the military (Machlis and Hanson 2008), Büscher and Fletcher (2018) continue the trajectory of 'green violence' with 'green wars', justifying the term because conservation is increasingly being conceived as a 'war' by its proponents, which signifies 'the most recent iteration in the historical intensification of power'. The authors of green militarization, violence and war recognize the violent colonial histories and forms of coercive conservation (see Peluso 1993) preceding them, yet see an intensification and onslaught of various forms of violence justified by conservation.

The Political Ecology of Counterinsurgency

Another line of inquiry has been through the lens of counterinsurgency. There are roughly two foundational schools of thought in counterinsurgency: The French and The British. The French approach was noticeably more forceful and less diplomatic, yet has taken on increasing mutations, while the British approach was known for articulating 'the idea of "minimum force" and reliance where possible on conventional civil and policing structures' (Rich and Duyvesteyn 2012: 9–10). Counterinsurgency emerges out of colonial warfare, and was developed as a doctrine to confront decolonization. Harry Truman's 1949[2] US president inaugural address, dripping with paternalistic and imperial overtones, not only introduced development, but also—and arguably—counterinsurgency as a methodology to win the 'hearts' and 'minds' of recalcitrant overseas populations.

Mao's rural insurrection in China made counter-guerrilla operations pressing, unsurprisingly leading to a community development pilot program in India—'the underbelly of China' (Berman 1983: 56)—that same year. Community development was expanded in 1952 with a $50 million 'package program' from USAID and the Ford Foundation to provide a culturally friendly model of modernization to address 'the village as the opponent of the state, the final bastion of habits and attitudes obstructing the smooth functioning of central power' (Cullather 2006, 2013 [2010]: 79).[3] Walt Rostow (1960), the father of mainstream development theory and national security planner, not only was an advocate of counterinsurgency (Halberstam 1972), but embodied its ethos by desiring to spread development through the enchantments of consumerism—by putting a 'television sets in the thatch hutches of the world'—and not shirking from carpet bombing—or the 'de-modernization'—of the recalcitrant (communist) enemy (Milne 2007; Cullather 2013 [2010]: 161). While Rostow has been wrongfully attributed to coining the term (see Khalili 2015), Captain Edward Lansdale's disciple, Lieutenant Colonel Sam Wilson, tasked with designing the first course on fighting guerrillas in the Special Warfare School, would be the person who coined the term and designed the first classes on 'counter-insurgency' in 1959 (Boot 2018). Counterinsurgency would not only entail violent repression, but as Lansdale (see Boot 2018: 319) himself explained, would entail undertaking 'missions of public works, welfare, health and education, as well as national security'.

[2] Truman H. (1949) Inaugural Address. Available at: http://www.presidency.ucsb.edu/ws/?pid=13282.

[3] These programs, Cullather (2013 [2010]) shows, later interlaced with other technologies for curbing 'red' revolution through the 'green' revolution, making the latter, to a certain degree, a technology of pacification.

Investigating the Police in the United States, Kristian Williams (2007 [2004]: 218) offered the simple formula: 'Community Policing + Militarization = Counter-insurgency'. Williams' (2007) research, later confirmed by the RAND Corporation (Gompert et al. 2008), was instrumental in demonstrating how the techniques of colonial warfare were taking hold domestically in the 1970s and have continued ever since (see Graham 2011; Williams et al. 2013). Counterinsurgency is a type of war—'low-intensity' or 'asymmetrical' combat—and style of warfare that emphasizes intelligence networks, psychological operations, media manipulation, security provision and social development to maintain governmental and extractive legitimacy (FM3-24 2014; Dunlap 2018a [2017]). Counterinsurgency is social warfare (Dunlap 2014a, 2019a), combining the brute force of 'hard' conventional warfare and 'soft' strategies that form a larger mutually reinforcing governmental-corporate strategy to discipline, enchant and engineer the 'hearts' and 'minds' of target populations. This frequently includes the pre-emptive and systematic targeting of non-violent protesters to prevent disruptions of controversial political and economic processes (Dunlap 2014b, 2016, 2018a).

Insights from land control and territorialization literature is complemented by political ecology research into counterinsurgency. Offering a historical overview of 'national natures' in Southeast Asia, Peluso and Vandergeest (2011: 603) highlight counterinsurgency techniques as instrumental not only to territorialization and nation state formation, but the creation of 'national forests', cities and the production of space generally. This recognition of counterinsurgency as foundational to producing space, subjectivities, securing land and governmental systems has inaugurated the political ecology of counterinsurgency. Counterinsurgency and economic growth retain a profound affinity in how military-security efforts create the conditions for capital accumulation, whether relating to conservation parks (Ybarra 2012, 2017; Marijnen and Verweijen 2016; Verweijen and Marijnen 2018), eco-tourisms (Devine 2014), development schemes (Copeland 2012; Grajales 2013; Paley 2014; Price 2014; Marijnen 2017), the green economy or 'climate change commodities' (Dunlap and Fairhead 2014: 938; Dunlap 2018a). While counterinsurgency is present in the sites covered, introducing the concepts of green militarization, violence and wars, there have been few authors from these camps to acknowledge the complementary nature of counterinsurgency in these sites of green militarization.

Counterinsurgency thus emerges as the popular methodology, or toolbox of repressive techniques, to suppress violent contestations, manage and engineer populations. Responding to the political reactions 'from below' literature, Dunlap (forthcoming) proposes a topology of political reactions 'from above' based on three case studies spanning wind energy development, coal mining and copper mining. Drawing from the political ecology of counterinsurgency, moreover, Dunlap (forthcoming) offers two overarching and mutually reinforcing categories: 'hard' coercive techniques and 'soft' social technologies of pacification. The 'hard' coercive techniques comprised of *overt coercion*—direct violence from police, military and extra-judicial forces—and *covert coercion*—extra-judicial operations emphasizing robberies, home invasions and murders in both private and public space. This is complemented by *overt surveillance*—police and extra-judicial actors blatantly watching homes, tailing individuals, leaving messages, flying helicopters/drones over homes or leaving messages on computers and cellphones after they have been hacked—and *covert surveillance*—taking on extra-judicial character with robberies, home invasions and murders in both private and public space.

On the other hand, there are six different, and again reinforcing, soft social technologies. The first, *social development*, does not enforce, but demonstrates the benefits of collaborating with governments, corporations and elites with the strategic deployment of social development: repairing irrigation and water infrastructure, church restoration programs, sponsoring schools, agricultural/livestock programs, medical clinics and so on. Second, *public relations* entails the deployment of print, radio and television advertisements; 'community' information centers; door-to-door or public canvasing; sponsoring festivals and cultural events; donating and/or pay rolling politicians, civil servants, police and landowners. Third, *scientific knowledge* is the employment of social (and natural) scientists in devising strategies to delegitimize or counter social movements, measure the effectiveness of social development interventions (see Price 2014) and conduct social/economic/environmental impact studies that favor the project or, as Stuart Kirsch (2014) calls it, 'corporate science'. Fourth, *counter-mobilizations* are the manufacturing or cultivation of groups to counter popular concerns, protests or insurrectionary tensions surrounding land grabs. This represents the ethos of counterinsurgency—to counter—emblematic of the 'it takes a thief to catch a thief' principle (Boot 2018: 321), which leads to the creation of counter-farmer and fisher people groups to discredit and fight other farmers and fisher peoples resisting enclosure, control and extraction.[4] Fifth, *green washing* is the deployment of environmental initiatives to further legitimize and/or brand a company, open new green economic markets and undermine environmentalists opposing the development project. Green washing has two principle modalities: *the project itself*—as with biofuels, conservation or wind energy development—and environmental 'offsetting' schemes that seek to compensate for extractivist damages. Finally, *infrastructure* can serve at least three pacification purposes: (1) as a method of controlling space as associated with infrastructural violence, (2) enclosing and countering land defender occupations/camps (see Brock and Dunlap 2018) and (3) non-material bureaucratic infrastructures akin to various types of consultations, including (faulty) free, prior and informed consent (FPIC) consultations (see Dunlap 2018c; Franco 2014).

These 'soft' social technologies of pacification can also combine with micro-policing techniques. Overlooked in the above mentioned literature, has been Michael Pendleton's (1998) outlining of four 'soft' law enforcement techniques used to enforce national parks in British Colombia, Canada (and likely elsewhere). *Encouraging compliance* employs symbolic violence centered around the 'display of the symbols of enforcement:' marked trucks, uniforms, signs; *threat enforcement* that threatened with 'hard' consequences; *non-enforcement* is the non-enforcement of laws when it benefited park rangers both energetically and economically, which also served as a method of information gathering from the person receiving leniency. Finally, *covert enforcement* utilized informants to investigate park violations. While there are clear over laps with the topology above, it also displays the detailed and subtle logic of policing conservation parks and contested land deals that are relationships and techniques that accumulate in many localities across the earth, embodying—in many instances—the virus of the Worldeater. Violence, counterinsurgency and the logics of war are both instrumental and carved into the process of land control and territorialization. Combined, these processes form the social engineering of extraction (Verweijen and Dunlap forthcoming). Social engineering, Christopher Hadnagy (2011: 10) defines as 'the act of manipulating a person to take an action that may or may not be in the "target's" best interest. This may include obtaining information, gaining access, or getting the target to take certain action'. The myriad of hard

[4] Nandini Sundar (2016) details these violent social technologies of counterinsurgency in the context of India's Naxalite insurgency on Indigenous territory.

coercive techniques and social technologies animate the social engineering of extraction, reinforcing the importance of the label land *grabbing*, meanwhile placing the planet, and its inhabitants, on an industrial trajectory that may not be in their best interest. Said differently, it structures the trajectory of the Worldeater.

Conclusion

The teeth and claws of the Worldeater, we have shown in this chapter, are many, powerful and adapting to new circumstances and political reactions. While the Worldeater's ripping and shredding continues, it sometimes happens in plain sight, sometimes stealthily in the dark, and often done with gusto by humans that believe in the war to be fought, to advance the nation or extractivism and security. This chapter offers an inventory of violent technologies of extraction that require acknowledgment, but also careful scrutiny and development in future research. We need to study how and when the teeth and claws do their work—a task that has been undertaken in the research examining imperial relationships in general, but in more detail in the works focusing on land control and territorialization. We see here how natural resource extraction is made acceptable or *livable* to humans as the social engineering of extraction proceeds apace as the political ecology of counterinsurgency animates and locates the politico-historical techniques of violence employed.

The Worldeater—techno-capitalist progress—exists and subsists on war, violence and trauma, as Perlman (2010) taught us. If there is one thing the long techno-capitalist trajectory—with its sequence of worms, octopuses and Leviathanic beasts—teaches us, it is that we should not underestimate the cunning, shapeshifting and ever-evolving ability to devise new violent technologies that not only repress human agency, but also possess it. Social engineering is a technical term for possessing human agency. The Worldeater is a conversation of possession, addiction, dependency and blindness that are accomplished through civil-military interventions (or so-called security strategies, see Bachmann et al. 2015) and solidified by politics. Hence, Patrick Wolfe's (2006: 388) now famous assertion: 'invasion is a structure not an event'. Existing inventories morph, new tools are invented, older ones converted and transformed—as this chapter has shown in the context of the 'green' economy. Or, in our preferred terms, the Worldeater sharpens its teeth and claws at any and all opposition to fulfilling its rapacious hunger for consuming human and non-human natures. This intensifies at the current conjuncture as extractivism reveals its totalizing tensions. Chapter 5 proceeds to uncover how the drive toward total extractivism increasingly blurs the lines between conventional and 'green' forms of extraction, revealing the outline of the systemically colonizing grid of the Worldeater.

Bibliography

Anthony, P. D. (2001 [1977]). *The Ideology of Work*. Oxon: Tavistock Press/ Routledge.

Bachmann, J., Bell, C., & Holmqvist, C. (2015). *War, Police and Assemblages of Intervention*. Oxon: Routledge.

Baka, J. (2013). The Political Construction of Wasteland: Governmentality, Land Acquisition and Social Inequality in South India. *Development and Change, 44*(2), 409–428.

Bebbington, A. (2012). *Social Conflict, Economic Development and the Extractive Industry: Evidence from South America*. London: Routledge.

Benjaminsen, T. A., & Bryceson, I. (2012). Conservation, Green/Blue Grabbing and Accumulation by Dispossession in Tanzania. *Journal of Peasant Studies, 39*(2), 335–355.

Berman, E. (1983). *The Ideology of Philanthropy: The Influence of the Carnegie, Ford, and Rockefeller Foundations on American Foreign Policy*. Albany, NY: State University of New York Press.

Boot, M. (2018). *The Road Not Taken: Edward Lansdale and the American Tragedy in Vietnam*. New York: Liveright Publishing.

Borras, S. M., McMichael, P., & Scoones, I. (2010). The Politics of Biofuels, Land and Agrarian Change: Editors' Introduction. *Journal of Peasant Studies, 37*, 575–592.

Bourdieu, P. (2001). *Masculine Domination*. Cambridge: Polity Press.

Bourgois, P. (2001). The Power of Violence in War and Peace: Post-Cold War Lessons from El Salvador. *Ethnography, 2*, 5–34.

Brock, A., & Dunlap, A. (2018). Normalising Corporate Counterinsurgency: Engineering Consent, Managing Resistance and Greening Destruction Around the Hambach Coal Mine and Beyond. *Political Geography, 62*, 33–47.

Büscher, B., & Fletcher, R. (2018). Under Pressure: Conceptualising Political Ecologies of Green Wars. *Conservation and Society, 16*, 105–113.

Büscher, B., & Ramutsindela, M. (2016). Green Violence: Rhino Poaching and the War to Save Southern Africa's Peace Parks. *African Affairs, 115*(458), 1–22.

Castro-Gomez, S. (2002). The Social Sciences, Epistemic Violence, and the Problem of the "Invention of the Other". *Nepantla: Views from South, 3*, 269–285.

Copeland, N. (2012). Greening the Counterinsurgency: The Deceptive Effects of Guatemala's Rural Development Plan of 1970. *Development and Change, 43*, 975–998.

Correa-Cabrera, G. (2017). *Los Zetas Inc.: Criminal Corporations, Energy, and Civil War in Mexico*. Austin, TX: University of Texas Press.

Cramer, C., & Richards, P. (2011). Violence and War in Agrarian Perspective. *Journal of Agrarian Change, 11*, 277–297.

Crosby, A. C., & Monaghan, J. (2018). *Policing Indigenous Movements: Dissent and the Security State*. Black Point: Fernwood Publishing.

Cullather, N. (2006). "The Target Is the People": Representations of the Village in Modernization and U.S. National Security Doctrine. *Cultural Politics, 2*, 29–48.

Cullather, N. (2013 [2010]). *The Hungry World: America's Cold War Battle Against Poverty in Asia*. Cambridge: Harvard University Press.

Devine, J. (2014). Counterinsurgency Ecotourism in Guatemala's Maya Biosphere Reserve. *Environment and Planning D: Society and Space, 32*, 984–1001.

Downey, L., Bonds, E., & Clark, K. (2010). Natural Resource Extraction, Armed Violence, and Environmental Degradation. *Organization Environment, 23*, 453–474.

Dunlap, A. (2014a). Power: Foucault, Dugger and Social Warfare. In B. Collective (Ed.), *The BASTARD Chronicles: Social War* (pp. 55–106). Berkeley, CA: Ardent Press.

Dunlap, A. (2014b). Permanent War: Grids, Boomerangs, and Counterinsurgency. *Anarchist Studies, 22*, 55–79.

Dunlap, A. (2016). Counter-Insurgency: Let's Remember Where Prevention Comes from and Its Implications. *Critical Studies on Terrorism, 9*, 380–384.

Dunlap, A. (2017). Wind Energy: Toward a "Sustainable Violence" in Oaxaca, Mexico. *NACLA, 49*, 483–488.

Dunlap, A. (2018a). Counterinsurgency for Wind Energy: The Bíi Hioxo Wind Park in Juchitán, Mexico. *The Journal of Peasant Studies, 45*, 630–652.

Dunlap, A. (2018b). The 'Solution' Is Now the 'Problem': Wind Energy, Colonization and the 'Genocide-Ecocide Nexus' in the Isthmus of Tehuantepec, Oaxaca. *The International Journal of Human Rights, 42*, 550–573.

Dunlap, A. (2018c). "A Bureaucratic Trap": Free, Prior and Informed Consent (FPIC) and Wind Energy Development in Juchitán, Mexico. *Capitalism Nature Socialism, 29*, 88–108.

Dunlap, A. (2019a). 'Agro sí, mina NO!' The Tía Maria Copper Mine, State Terrorism and Social War by Every Means in the Tambo Valley, Peru. *Political Geography, 71*, 10–25.

Dunlap, A. (2019b). *Renewing Destruction: Wind Energy Development in Oaxaca, Mexico*. London: Rowman & Littlefield.

Dunlap, A. (forthcoming). Wind, Coal and Copper: The Politics of Land Grabbing, Counterinsurgency and the Social Engineering of Extraction. *Globalizations*.

Dunlap, A., & Fairhead, J. (2014). The Militarisation and Marketisation of Nature: An Alternative Lens to 'Climate-Conflict'. *Geopolitics, 19*, 937–961.

Elden, S. (2010). Land, Terrain, Territory. *Progress in Human Geography, 34*(6), 799–817.

Eldridge, E. R., & Reinke, A. J. (2018). Introduction: Ethnographic Engagement with Bureaucratic Violence. *Conflict and Society, 4*, 94–98.

Fairhead, J., Leach, M., & Scoones, I. (2012). Green Grabbing: A New Appropriation of Nature? *Journal of Peasant Studies, 39*, 237–261.

Federici, S. (2004). *Caliban and the Witch: Women, the Body and Primitive Accumulation*. New York: Autonomedia.

Finley-Brook, M., & Thomas, C. (2011). Renewable Energy and Human Rights Violations: Illustrative Cases from Indigenous Territories in Panama. *Annals of the Association of American Geographers, 101*, 863–872.

FM3-24. (2014). *Insurgencies and Countering Insurgencies*. Retrieved from http://fas.org/irp/doddir/army/fm3-24.pdf.

Foucault, M. (1995 [1977]). *Discipline and Punish: The Birth of the Prison*. New York: Random House.

Foucault, M. (1998 [1978]). *The Will to Knowledge: The History of Sexuality: 1*. London: Penguin Books.

Foucault, M. (2003 [1997]). *"Society Must Be Defended": Lectures at the College de France 1975–1976*. New York: Picador.

Foucault, M. (2007 [1978]). *Security, Territory, Population: Lectures at the College de France 1977–1978*. New York: Picador.

Franco, J. (2014). Reclaiming Free Prior and Informed Consent (FPIC) in the Context of Global Land Grabs. Transnational Institute for Hand off the Land Alliance.

Galeano, E. (1997 [1973]). *Open Veins of Latin America: Five Centuries of the Pillage of a Continent*. London: Monthly Review Press.

Galtung, J. (1969). Violence, Peace, and Peace Research. *Journal of Peace Research, 6*, 167–191.

Gelderloos, P. (2013). *The Failure of Nonviolence: From Arab Spring to Occupy*. Seattle, WA: Left Bank Books.

Gompert, D. C., Grissom, J. G., IV, et al. (2008). *War by Other Means: Building Complete and Balanced Capabilities for Counterinsurgency*. RAND Counterinsurgency Study Final Report. Santa Monica RAND Corporation National Security Research Division.

Graham, S. (2011). *Cities Under Siege: The New Military Urbanism*. New York: Verso.

Grajales, J. (2013). State Involvement, Land Grabbing and Counter-Insurgency. *Development and Change, 44*, 211–232.

Hadnagy, C. (2011). *Social Engineering: The Art of Human Hacking*. Indianapolis, IN: Wiley Publishing.

Halberstam, D. (1972 [1969]). *The Best and the Brightest*. New York: Random House.

Holmes, G. (2014). What Is a Land Grab? Exploring Green Grabs, Conservation, and Private Protected Areas in Southern Chile. *Journal of Peasant Studies, 41*(4), 547–556.

Khalili, L. (2015). Counterterrorism and Counterinsurgency in the Neoliberal Age. In A. Ghazal & J. Hanssen (Eds.), *Oxford Handbook of Contemporary Middle-Eastern and North African History*. Oxford: Oxford University Press.

Kirsch, S. (2014). *Mining Capitalism: The Relationship Between Corporations and Their Critics*. Berkeley, CA: University of California Press.

Kleinman, A. (2000). The Violences of Everyday Life. In V. Das, A. Kleinman, M. Ramphele, et al. (Eds.), *Violence and Subjectivity* (pp. 226–241). Berkeley, CA: University of California Press.

Li, T. M. (2018). After the Land Grab: Infrastructural Violence and the "Mafia System" in Indonesia's Oil Palm Plantation Zones. *Geoforum, 96*, 328–337.

Lund, C. (2016). Rule and Rupture: State Formation Through the Production of Property and Citizenship. *Development and Change, 47*, 1199–1228.

Lunstrum, E. (2014). Green Militarization: Anti-Poaching Efforts and the Spatial Contours of Kruger National Park. *Annals of the Association of American Geographers, 104*, 816–832.

Machlis, G. E., & Hanson, T. (2008). Warfare Ecology. *BioScience, 58*(8), 729–736.

Marijnen, E. (2017). The 'Green Militarisation' of Development Aid: The European Commission and the Virunga National Park, DR Congo. *Third World Quarterly, 38*, 1566–1582.

Marijnen, E., & Verweijen, J. (2016). Selling Green Militarization: The Discursive (Re)Production of Militarized Conservation in the Virunga National Park, Democratic Republic of the Congo. *Geoforum, 75*, 274–285.

Marker, M. (2003). Indigenous Voice, Community, and Epistemic Violence: The Ethnographer's "Interests" and What "Interests" the Ethnographer. *International Journal of Qualitative Studies in Education, 16*, 361–375.

Marx, K. (1982 [1867]). *Capital: A Critique of Political Economy Vol. I*. London: Penguin Books.

Massé, F., & Lunstrum, E. (2015). Accumulation by Securitization: Commercial Poaching, Neoliberal Conservation, and the Creation of New Wildlife Frontiers. *Geoforum, 69*, 227–237.

Milne, D. (2007). "Our Equivalent of Guerrilla Warfare:" Walt Rostow and the Bombing of North Vietnam, 1961–1968. *The Journal of Military History, 71*, 169–203.

Moses, A. D. (2008). *Empire, Colony, Genocide: Conquest, Occupation, and Subaltern Resistance in World History* (*War and Genocide*). Oxford: Berghahn.

Moses, D., & Stone, D. (2013). *Colonialism and Genocide*. London: Routledge.

Nixon, R. (2011). *Slow Violence and the Environmentalism of the Poor*. Cambridge: Harvard University Press.

O'Brochta, W. (2019). A Meta-Analysis of Natural Resources and Conflict. *Research & Politics, 6*(1). https://doi.org/10.1177/2053168018818232.

Ojeda, D. (2012). Green Pretexts: Ecotourism, Neoliberal Conservation and Land Grabbing in Tayrona National Natural Park, Colombia. *The Journal of Peasant Studies, 39*, 357–375.

Paley, D. (2014). *Drug War Capitalism*. Oakland, CA: AK Press.

Peluso, N. L. (1993). Coercing Conservation? The Politics of State Resource Control. *Global Environmental Change, 3*, 199–217.

Peluso, N. L., & Lund, C. (2011). New Frontiers of Land Control: Introduction. *The Journal of Peasant Studies, 38*, 667–681.

Peluso, N. L., & Vandergeest, P. (2011). Political Ecologies of War and Forests: Counterinsurgencies and the Making of National Natures. *Annals of the Association of American Geographers, 101*, 587–608.

Peluso, N. L., & Watts, M. (2001). *Violent Environments*. Ithaca, NY: Cornell University Press.

Pendleton, M. R. (1998). Policing the Park: Understanding Soft Enforcement. *Journal of Leisure Research, 30*(4), 552–571. https://doi.org/10.1080/00222216.1998.11949847.

Perelman, M. (2007). Primitive Accumulation from Feudalism to Neoliberalism. *Capitalism Nature Socialism, 18*, 44–61.

Perlman, F. (2010 [1983]). *Against His-Story, Against Leviathan*. Detroit: Red and Black Press.

Polanyi, K. (2001 [1944]). *The Great Transformation: The Political and Economic Origins of Our Time*. Boston: Beacon Press.

Price, D. H. (2014). Counterinsurgency by Other Names: Complicating Humanitarian Applied Anthropology in Current, Former, and Future War Zones. *Human Organization, 73*, 95–105.

Rasmussen, M. B., & Lund, C. (2018). Reconfiguring Frontier Spaces: The Territorialization of Resource Control. *World Development, 101*, 388–399.

Rich, P. B., & Duyvesteyn, I. (2012). The Study of Insurgency and Counterinsurgency. In *The Routledge Handbook of Insurgency and Counterinsurgency* (pp. 1–19). London: Routledge.

Rignall, K. E. (2016). Solar Power, State Power, and the Politics of Energy Transition in Pre-Saharan Morocco. *Environment and Planning A, 48*(3), 540–557.

Rodgers, D., & O'Neill, B. (2012). Infrastructural Violence: Introduction to the Special Issue. *Ethnography, 13*, 401–412.

Rodney, W. (2009 [1972]). *How Europe Underdeveloped Africa*. Washington, DC: Howard University Press.

Rostow, W. (1960). *The Stages of Economic Growth: A Non-Communist Manifesto*. Cambridge: Cambridge University Press.

Scheper-Hughes, N. (1992). *Death Without Weeping. The Violence of Everyday Life in Brazil*. Berkeley; Los Angeles; Oxford: University of California Press.

Scott, J. C. (1998). *Seeing Like a State: How Certain Schemes to Improve the Human Condition Have Failed*. New Haven, CT: Yale University Press.

Short, D. (2016). *Redefining Genocide: Settler Colonialism, Social Death and Ecocide*. London: Zed Books.

Siamanta, Z. C. (2017). Building a Green Economy of Low Carbon: The Greek Post-Crisis Experience of Photovoltaics and Financial 'Green Grabbing'. *Journal of Political Ecology, 24*, 258–276.

Siamanta, Z. C. (2019). Wind Parks in Post-Crisis Greece: Neoliberalisation Vis-à-Vis Green Grabbing. *Environment and Planning E: Nature and Space*. https://doi.org/10.1177/2514848619835156.

Spivak, G. (1988). Can the Subaltern Speak? In C. Nelson & L. Grossberg (Eds.), *Marxism and the Interpretation of Culture* (pp. 271–313). Urbana: University of Illinois Press.

Springer, S., & Le Billon, P. (2016). Violence and Space: An Introduction to the Geographies of Violence. *Political Geography, 52*, 1–3.

Stephen, L. (2000). The Construction of Indigenous Suspects: Militarization and the Gendered and Ethnic Dynamics of Human Rights Abuses in Southern Mexico. *American Ethnologist, 26*, 822–842.

Sundar, N. (2016). *The Burning Forest: India's War in Bastar*. New Delhi: Juggernaut.

Thompson, E. P. (1991 [1963]). *The Making of the English Working Class*. London: Penguin Books.

Verweijen, J., & Dunlap, A. (forthcoming). The Evolving Techniques of Social Engineering, Land Control and Managing Protest Against Extractivism: Introducing Political (Re)Actions 'From Above'. *Political Geography*.

Verweijen, J., & Marijnen, E. (2018). The Counterinsurgency/Conservation Nexus: Guerrilla Livelihoods and the Dynamics of Conflict and Violence in the Virunga National Park, Democratic Republic of the Congo. *The Journal of Peasant Studies, 45*, 300–320.

Vidal, J. (2008). *The Great Green Land Grab*. Retrieved from http://www.theguardian.com/environment/2008/feb/13/conservation.

Williams, K. (2007 [2004]). *Our Enemies in Blue: Police and Power in America*. Cambridge: South End Press.

Williams, K., Munger, W., & Messersmith-Glavin, L. (2013). *Life During Wartime: Resisting Counterinsurgency*. Edinburgh: AK Press.

Wolfe, P. (2006). Settler Colonialism and the Elimination of the Native. *Journal of Genocide Research, 8*, 387–409.

Ybarra, M. (2012). Taming the Jungle, Saving the Maya Forest: Sedimented Counterinsurgency Practices in Contemporary Guatemalan Conservation. *The Journal of Peasant Studies, 39*, 479–502.

Ybarra, M. (2017). *Green Wars: Conservation and Decolonization in the Maya Forest*. Oakland, CA: University of California Press.

CHAPTER 5

The Worldeater(s) in Process: Uncovering the Nexus of Conventional and 'Green' Extraction

Abstract This chapter interrogates the subtle shifts and blurring lines between conventional extraction—mineral and hydrocarbon—and 'green' extraction—intensive agriculture and renewable energy. Through the careful assembly of extensive amounts of empirics straddling these modalities of extraction, we identify and uncover a crucial *nexus*. We argue that this nexus is key in animating the present imperative of total extractivism. The nexus further reveals the violent technologies of extraction at work as it lays out further dimensions of the organization of the rapaciously devouring machinery spreading its grid across ever-increasing portions of the earth. The chapter thus traces the infrastructure—the 'body' of the Worldeater(s)—through mines, plantations, factory farms and renewable energy to chart the formation and/or spread of the Worldeater.

Keywords Conventional extraction • Green extraction • Extractivism • Infrastructure • Renewable energy

A. Dunlap, J. Jakobsen, *The Violent Technologies of Extraction*,
https://doi.org/10.1007/978-3-030-26852-7_5

Renewable resources were the conventional resource for civilizations and, later, state formation. It was not until the nineteenth century that coal, followed by other hydrocarbon resources, would take center stage as the conventional fuel and driving force behind capitalism and industrialism (Malm 2016). While mining is an ancient practice of both civilizations and Indigenous societies (see Lahiri-Dutt 2018b), the latter was on the micro-scale and embodied different socio-ontological relationships. Taussig (1980), as we have seen, reveals the severity of the rupture as capitalist mining and plantations descend on Indigenous and peasant societies—converting them into 'neophyte proletarians'—and causing a maelstrom of spiritual, social and ecological disruptions. It is through such violent maelstroms that rural areas are incorporated in—devoured by—the machinery of the Worldeater, converting rural societies, non-humans and natural environments into 'entrails' of the Leviathanic beast. Taussig's work thus alerts us to the intertwinement of otherwise seemingly disparate modalities of extraction—mines and plantations—the similarity of their function and spiritual-politico-economic tensions. This chapter turns to interrogate the subtle shifts and blurring lines between conventional extraction—mineral and hydrocarbon—and 'green' extraction—intensive agriculture and renewable energy. Through the careful assembly of extensive amounts of empirics straddling these modalities of extraction, we uncover a crucial *nexus*. We argue that this nexus is key in animating the present imperative of total extractivism. The nexus, in other words, reveals the violent technologies of "green" and conventional extraction at work as it organizes the rapaciously devouring machinery that spreads the Worldeater's grid across ever-increasing portions of the earth.

Total Extractivism: The Trajectory of Progress

Focusing on the Global South—and Latin America in particular—extractivism tends to be understood as a staple feature of colonial, neocolonial and, increasingly, neoliberal capitalist economies that, according to Eduardo Gudynas (2013 [2011]), retain three features: extract large quantities of material; cause large-scale degradation of ecosystems; and export raw materials with minimal processing. The overall destructive character of extractivism is obvious, not limited to ecosystems but also changes local economies. Changes occur with the arrival of outsiders, their new habits and their machines, but also with failing to deliver the levels of employment, revenues and social development once promised by companies and their corporate social responsibility schemes (see Bebbington 2012; Kirsch 2014; Gamu and Dauvergne 2018). Instead, colonial and capitalist extractivism has led to extensive critique, protest, resistance and assassination (Middeldorp et al. 2016; Birss 2017; Dunlap 2019a). The onset of leftist governments in Latin America in the twenty-first century—the 'Pink Tide'—has led to new terms and acceleration of extraction in the region (Vergara-Camus and Kay 2018). Dubbed 'neo-extractivism' by Gudynas (2009), where leftist governments justified continued extractivism on the basis of a development policy concerned with wealth distribution and industrialization, which has been considered a failure on social and environmental grounds (Acosta 2013; Rosales 2016; Vergara-Camus and Kay 2018), consequently serving the same neoliberal wine in new state-centered bottles labeled as 'post-neoliberalism' (see Vela-Almeida 2018). In reality, Petras and Veltmeyer (2014) argue, this has all enabled the intensification of accumulation by dispossession (Harvey 2003) robbing both human and non-human populations alike. Extractivism, in other words, consumes human and non-human resources, 'leaving a void' (Ye et al. 2019). Mining vitality from the earth until nothing is left but a void—Worldeater excretion—extractivism displays the rapacious appetite of the global capitalist economy as it moves across the earth in search of new sources of vitality. Indeed, the wormlike character of extractivism is noted—though not traced further, leaving the reader pondering its deeper implications—by Ye and colleagues (2019: 5):

> Extractivism represents a politico-economic formation that is constantly on the move: on the one hand, it is feeding itself by ruthlessly exploiting (and depleting) the resources it controls; on the other, it is grabbing new resources in order to continue its operation. Ongoing conquest is a material need here: instead of reproducing resources, new ones need to be conquered.

We have already established the variegated force, intensity and velocity by which the Worldeater's unremitting conquest—*worldeating*—unfolds. This view of techno-capitalist industrialism appears to leave questions of 'impact' redundant, mere pedantry. Lo and behold, we find what we expect: void-making—devouring of vital life and degradation left to heal in its wake. These operations, however, are not confined to conventional extraction. Neo-extractivism, as it emerged alongside the green economy and climate change mitigation legislation, came to include new forms of 'green extractivism'.

Greening Extractivism

While it is safe to say that 'green' extractivism—the extraction of renewable resources (in the broadest sense) from the earth—is nothing new, it has gained center stage in the era of the greening Capitalocene (see Bonneuil and Fressoz 2016; Moore 2015). Not the least, the industrialization of agriculture in general turns it into what Richard Walker (2004: 191) calls 'petrofarming' whose '(il)logical end' is 'to eliminate the land as the basis of production' as it exhausts the earth, exchanging nutrients with chemicals, exchanging natural circuits with 'offsets' and externalities (see also Jakobsen 2018). Already conventional and green extractivism begin to merge into the same process of exhaustion, where petrofarming—and other forms of extraction—causes cascading socio-ecological damage—toxification, water contamination, greenhouse gas emission, to name a few. Meanwhile, un- or under-accounting for the damage done is standard protocol (Weis 2010). Calling these processes 'extraction' does not require any great stretch of the imagination, although we may add further ingredients in order to justifiably term modalities of industrialized agriculture as a form of extractivism. 'Agricultural extractivism', popularly described by Gudynas (2010: 2), refers to the intensification of monocrops for the exportation of an agricultural commodity, employing transgenic crops, mechanization and chemical inputs. Agricultural extractivism is employed by various authors (Svampa 2013; Giarracca and Teubal 2014; Petras and Veltmeyer 2014). Referring generally to 'large-scale, intensive monocrop production for export', Mckay (2017: 202) asks: what is the extractive character of agrarian extractivism (as opposed to industrial agriculture)? The variables to qualify extraction are many; following Bernstein (2010), we find that specific articulations of 'land control and use, labor relations, surplus distribution, and the social relations of consumption, reproduction and accumulation' are central to extraction. Accounting for these variables, based on variegated scales of horizontal and vertical distribution, Mckay (2017: 202) claims not all agribusiness 'is inherently extractive as such', leading him to offer four inter-linked features of agrarian extractivism:

1. Large volumes of materials extracted destined for export with little or no processing;
2. Value-chain concentration and sectoral disarticulation;
3. High intensity of environmental degradation;
4. Deterioration of labor opportunities and labor conditions in the areas/sector.

These criteria can act as a guide in measuring the actions of agribusiness engaging in novel forms of contracting, distribution and export, often subcontracting smallholders into corporate supply chains (see Fairbairn et al. 2014). The green or renewable nature of agrarian extractivism, however, is (again, expectedly) tenuous. Acosta (2013: 63) reminds us that as forest and soil fertility decline, the land risks becoming non-renewable, thereby blurring conventional and green extractivism into one. Indeed, the essence of extraction is the consumption, digestion and excretion of the earth's vitality—wind, water, sun, soil and all of its living inhabitants—into the body of the Worldeater.

On this background, this chapter—in Perlman's words—continues the description of the 'beasts' body'. It seeks to map, detail and scrutinize the organization of the systemically colonizing grid as it expands across—and into—the earth, covering a spectrum of violent technologies: from mines, to plantations, to factory farms and onward to renewable energy. While these examples show the spread of the Worldeater to be wide-ranging, we maintain that the Worldeater progresses at variegated intensities, sometimes more complete, other times with fewer layers and organs or centralized nodes, yet the progressive operation of worldeating continues. This chapter merely reveals parts of a vast body in formation. Guided by Perlman and Taussig's 'neophyte proletarians', we start with how the Worldeater possesses and devours rural people.

Entrails Making: 'Extractive Peasants' from Mines to Plantations

The imperative of total extractivism thrusts humans not only into the belly of the Beast, but also into the earth itself. Central to the rise of capitalism and the expansion of the Worldeater were mines (Moore 2010) and, later, plantations (Mintz 1985). Mines and plantations, we can say, were always 'core' and never 'periphery' to the global capitalist economy. Enticing vulnerable rural populations and smallholders with the promise of an income—survival and consumerism—in the face of the global 'corporate food regime' (McMichael 2013), the extractivist grid spreads by having peasants perform work in mines and plantations. Kuntala Lahiri-Dutt (2018a: 1) invites us to think of these processes of incorporation as the formation of 'extractive peasants'. These are

> rural workers in less affluent nations who have been shifting to, or alternating with, other livelihoods to make a living through commodity extraction. They work either independently or in groups; with a regional or family history, or as recent opportunistic migrants to a new site, extracting from small mineral deposits, with or without licence; as wage labourers on an irregular or contract basis for small mines and quarries; or steal from existing and abandoned mines. Their extractive practices are equally wide-ranging: from artisanal, highly sophisticated practices to working as wage labourers in licenced quarries that comprise part of the informal sector of the national economies and represent a broad array of capitalisation and labour conditions.

This is *entrails making* in action—devils, vampires and shapeshifting monstrous creatures follow in its wake. The notion of the extractive peasant, moreover, brings to view the differentiation in colonizing violent technologies. Departing from the 'hard' and 'soft' technologies discussed in the previous chapter, what is at play here is, crucially, the way the imperative of total extractivism feeds on neoliberal capitalist dynamics and 'adverse incorporation' (Borras and Franco 2013; Hall et al. 2015) into conventional and—as we will see—sometimes seamlessly green extractive industries. The extractive peasant extends peasant relationships to one or multiple commodity frontiers, which brings up and intertwines into a number of political debates.

There is the politics of the 'peasant'. Lahiri-Dutt (2018a: 8) explains: 'First, whether all peasants must necessarily be farmers; second, whether they must all be strictly attached to land; third, whether they participate in the market economy; and fourth, whether peasants experience change homogeneously as a single class'. While these questions reflect longstanding debates—recounted in Chap. 3—in critical agrarian studies, suffice to say that peasants traditionally have always been engaged in multiple crafts and work (see Peluso 2017). Easther Chigumira (2018), for example, argues that peasants in Zimbabwe engage in artisanal small-scale mining as a means of retaining or even strengthening their land-based economies. Peasants have, and continue to be as extractive peasants, attached to the land and, if anything, are adapting to 'the withdrawal of the state and entry of foreign capital' (Lahiri-Dutt 2018a: 7)—the movements and shape-making of the Worldeater. Third, peasants and small-scale miners have always negotiated various degrees of engagement with the market economy. Fourth, peasants are not homogeneous and are adapting to socio-economic circumstances imposed on them, while simultaneously 'informal mining is a deliberate production strategy to optimize their chances of moving out of dire poverty' (Lahiri-Dutt 2018a: 9). However, Lahiri-Dutt (2018a: 9) stresses that peasants shifting into or engaging in mining are not engaging in a 'survival strategy', seeking to resist the narrative of 'poor as victim' (see Escobar 2012), as they can always take action to survive via theft, begging, prostitution or, better, rebellion (see Dunlap 2018b). Yet, this real and empowering narrative should not sideline the social engineering of desire, extraction and, consequently, poverty outlined so well by Escobar (2012) and others post-development thinkers (Illich 1970; Rahnema and Bawtree 1997; Esteva 2014; Demaria and Kothari 2017). The state, economy and their imperatives are something inherited and reproduced in variegated ways and means, leaving us to wonder *who* is steering the trajectory of extractivism: 'the rich' (aka 'capitalists')? Politicians? Corporations? Consumer demand? Or is the trajectory steered, instead, by an entire systemic apparatus of self-reinforcing logics and implications operated by humans who are possessed by it and continue to operate it despite it destroying their habitats and trading their environment for commodities?

Extractive peasants' agency is important, especially when all small-scale mining—the most common form of mining, especially in the South—is considered illegal, 'black market' and environmentally degrading (compared to large-scale corporate mines), as it challenges corporate (and sometimes governmental) control over resources (Lahiri-Dutt 2018a). Yet, as Lahiri-Dutt (2018a: 4–5) acknowledges, the forced acceptance of artisanal mining through the persistence of the extractive peasant plays into a neoliberal strategy that opens up small-scale mining through subcontracting and 'complex supply chains and networks' similar to the way agricultural peasants are integrated into corporate agricultural subcontracting regimes (see White et al. 2012; Fairbairn et al. 2014). Relatedly, writing about the Bolivian context, Marston and Perreault (2017) argue for seeing hegemony arising in 'extractive regimes' through the integration of peasants into neoliberal governance arrangements via cooperatives. Nevertheless, there is the need for differentiated studies, as the specific form of mining shapes specific labor regimes. Whereas the 'extractive peasant' refers primarily to small-scale forms of mining, open pit mining 'requires the use of "advanced" labor-displacing technologies' (Giarracca and Teubal 2014: 48). As landscapes are dynamited and minerals excavated through large-scale techniques, the enrollment of personnel for operations—the means whereby humans come to perform worldeating operations, we may add—takes on other forms.

Acknowledging the relationships between plantations and mines in Indonesia, Peluso (2017) charts the mineral extractive activities of smallholders, finding 'extractive peasants' as well as 'small entrepreneurs', but in her view, 'all are smallholders' (Peluso 2017: 839). 'Smallholder' is employed to challenge the apolitical meaning of the term as it relates to the historical context of governmental strategies to manufacture the 'smallholder slot' to engage in various activities, such as small-scale farming, agroforestry, plantation production and, now, small-scale mining. The 'smallholder slot', Peluso (2017) says, was constructed and changed via commodity production, but also through state programs of territorial control. In the 1960s, she narrates, 'commercially oriented smallholders identified as "Chinese" were evicted' and the smallholder slot replaced with 'so-called indigenous "Indonesians"', thus socially constructing the smallholder through commodities and governmental programs. As with Lahiri-Dutt (2018a) above, the smallholder slot is responding to government policy—that eliminates agricultural subsidies and reconfigures food supply chains—and new imperatives for resource extraction development. The smallholder exercises agency within a particular socio-political framework, but this framework is governed by the state, capitalist relationships and the existential drive of the Worldeater.

Bringing more fully to view the violent technologies deployed through infrastructure, Tania Li (2018) argues for seeing plantations in Indonesia as forming what she calls a 'mafia system' where predatory actors engage in widespread rent seeking—a feature that we otherwise know as exemplary of the 'neo-extractivist' tendency. Predation intensifies and exceeds the mafia system, Li argues, as plantations draw in socio-ecological life in its totality in 'plantation zones', that is, distinct spaces organized around extraction. The violence exerted in, by and through the plantation zones is *infrastructural*—embedded in roads, fences, housing,[1] in a word: the grid of the Worldeater. This, we may add, links plantation systems to industrial systems of control more broadly such as prisons (McKittrick 2011) and leaves little doubt about the crucial importance of viral infrastructural spread for organizing and channeling the system of total extractivism, but in ways that can be diffuse and decentralized (as compared to the hyper-centralism postulated by Ye et al. 2019). Li (2018: 329–30) continues aptly to emphasize how this means that plantations—as violent technologies of extraction—cannot be considered otherwise, cannot be 'converted' into life-sustaining technologies:

> Indonesia's plantations are *routinely* violent because of the forms of life they destroy, the resources they monopolize, the futures they preclude, and the set of material, social and political relations they enable and fix in place. Since the violence is built-in, and intensifies as plantations multiply and expand, it is well-nigh impossible to revere engineer or retrofit plantations.

The entailment, of course, is that these violent technologies need to be subverted and decomposed. While Li only passingly mentions how the plantation is indeed an extractive practice, Perreault (2018: 2) replies to her paper that '[I]n a fundamental sense, both the plantation and the mine

[1] See Chap. 3 and Rodgers and O'Neill's (2012) on Infrastructural Violence.

remind us that capitalism is, unavoidably, an environmental project' of 'industrial-scale appropriation of natural resources in the service of accumulation' that 'reconfigure social relations and local ecologies'. Effectively bringing together what have often been considered the respective domains of CAS and political ecology: the plantation/mine figure serves as an illustration of the inextricable industrial logics underpinning both 'domains' of total extractivism.[2]

These interpenetrated logics come to the fore in industrial tree plantations (ITPs), frequently defined as 'large-scale monocultures of tree crops—mainly eucalypts, pines, rubber tree and oil palm—that are managed intensively, which generally involves the use of agrochemicals, cloned or genetically modified trees, and short rotations' (Gerber 2011: 165–6). Feeding into the global capitalist economy of interchangeable uses, 'flex trees' surge in importance across the South, commonly in close collaboration between states and companies (Kröger 2016). This expansion is highly conflictual: Looking at the expansion of such tree plantations in Chile, for example, we find plantation companies spreading their land control over vast territories with state backing. Strong resistance by smallholders has marked these expansions (González-Hidalgo and Zografos 2017). Indigenous groups and land defenders are commonly attacked by state, and paramilitary forces through 'hard' coercion and 'soft' greening and development techniques (Dunlap 2018a, 2019a; Brock and Dunlap 2018). Indeed, the tendency for Indigenous populations and smallholders to attempt the occupation of their territories, leading to brutally repressive violence, has been reported across the World (Gerber 2011; Crosby and Monaghan 2018; Dunlap 2018a). Yet, as Yunan Xu (2019) shows, Chinese ITPs can also be of smaller scale, owned by individual villagers. It is thus not necessarily productive to think of these expansions as exclusively driven by the agency of external agents, as we know from work on 'boom crops' elsewhere, driven significantly by local people's own actions (Hall 2011; Li 2014). While Perlman's 'armour' and 'masks' spread, designating top-down from the bottom up strategies (see Dunlap and Fairhead 2014), the Worldeater feeds on smallholders and others whose 'agency' is regimented into the trajectory of total extractivism.

[2] It is thus not coincidental that the 'game' of naming our current epoch of techno-capitalist industrial development includes not only the Capitalocene (Moore 2015) but also the 'Plantationocene' (Haraway 2015) among others.

These agro-extractivist logics bear resemblance to other extractive 'sectors' as well, including oil. Michael Watts (2004: 54) explains the political ecology of violence in Nigeria in strikingly relevant terms: 'Petro-capitalism', he explains, 'operates through a particular sort of "oil complex" (a unity of firm, state [and its security forces], and community) that is territorially constituted through oil concessions'. Mining complexes, plantation complexes, oil complexes: Total extractivism pushes forward and we still fail to discuss in greater depth the bioengineering and an array of technological development taking place through robotics, nanotechnology and, overall, artificial intelligence that arguably is manifesting collective consciousness of the Worldeater. While these areas, among others, deserve greater scrutiny, another example is the Animal-Industrial Complex.

Chained Commodities: Factory Farms and the Extraction of Vital Life

Non-human animals are among the oldest and most numerous of entrails in the womb of the Worldeater. While we do not rule out the ability of non-humans to resist the onslaught of worms and octopuses (see Hribal 2010), it is clear that their teeth and claws cannot compete in size with the Leviathanic beasts, and their abilities at subverting violent technologies are weaker than those of humans. The present situation, therefore, is one of catastrophic possession, loss and suffering for non-humans. While we have discussed the onslaught and spread of mines and plantations, another colonizing force of rural—and increasingly urban—space is factory farm infrastructure. Not only does this infrastructure spread across the world, but it is the lived embodiment of scientific violence and extermination. The factory farm represents a killing machine apparatus that has executed 1,485,986,756 pigs, 66,566,725,000 chickens, 26,435,897 buffaloes, 304,414,858 cattle, 3,011,798,000 ducks, 464,598,299 goats in 2017 alone (FAO 2018). Statistics likes these, as Richard White (2017: 274) recognizes, are virtually incomprehensible and benumbing for human minds. This amounts to the staggering number of 70 billion non-human animals being slaughtered every year, which is projected to continue increasing as human consumption of meat surges, amounting to the mind-boggling number of 120 billion non-human animals killed yearly by 2050 (Weis 2016).[3]

[3]These numbers do not include sea animals and thus 'do not illustrate the full scale of death' (Wadiwel 2015: 6).

Predicated upon techno-capitalist industrial logics, the concerted mass slaughter of non-humans often called—in sanitized language—'industrialized livestock production' contributes further to dissolving the conventional/green extraction binary. Factory farming extracts vitality from sentient beings for the purpose of capital accumulation—exchange value—and consumer cultures (Gunderson 2011). Like extractivisms uniformly, factory farms cause toxic pollution, health hazards, greenhouse gas emissions and cascading socio-environmental degradation (Weis 2013). In doing so, it turns non-human animals into what Bob Torres (2007: 11) calls 'chained commodities', that is, 'superexploited living commodities' whereby 'animals become nothing more than living machines, transformed from beings who live for themselves into beings that live for capital'. While the logic of capitalism is evident, this extraction of vitality has, as we have inferred in this book, older roots. David Nibert (2013: 12) argues at length that it can be traced to the historical onset of domestication, or what he calls 'domesecration', namely 'the systemic practice of violence in which social animals are enslaved and biologically manipulated, resulting in their objectification, subordination, and oppression'. The extensive interspecies violence, Nibert (2013) further argues, is intrinsically related to patterns of violence against other humans throughout history (see also White 2017): both enabling and promoting waves of violence as exploitation of the earth proceeds apace.

The extraction of vitality from chained commodities takes place, increasingly across the globe, within what Barbara Noske (1989) described as the 'Animal-Industrial Complex' that remains subject of a growing body of literature (White 2017; Weis 2013; Emel and Neo 2017; Jakobsen and Hansen 2019). The Animal-Industrial Complex, Richard Twine (2012: 23) summarizes, can be addressed as

> a partly opaque and multiple set of networks and relationships between the corporate (agricultural) sector, governments, and public and private science. With economic, cultural, social and affective dimensions it encompasses a range of practices, technologies, images, identities and markets.

Tony Weis (2013, 2016) similarly describes the expansion of industrial livestock as unfolding within 'the industrial grain-oilseed-livestock complex'. Introducing crops into a broadened agro-complex, Weis opens for exploring how agrarian extractivism in particular places (such as soy in Latin America) feed into an expanding human desire for meat in other places (such as China) (see Jakobsen and Hansen 2019). Indeed, Weis (2013: 8) goes on to argue that this complex

> is the dominant system of agriculture across the temperate world, and is spreading to significant parts of the tropics. Its landscapes can be likened to islands of concentrated livestock within seas of grain and oilseed monocultures, with soaring populations of a few livestock species reared in high densities, disarticulated from the surrounding fields.

We see clearly the mine/plantation/factory farm figure in its morphing forms. While factory farming is particularly apt here, one may hold that the capitalist logic of extracting vitality is retained also in 'free-range' or otherwise seemingly 'ethical' animal meat: these animals also end up killed for capitalist profit (White 2017; Stuart et al. 2013). We can extend this to 'humanitarian violence' and consultation procedures that offer the illusion of participation and decision making, both in the end result in geopolitical domination, land grabbing and resource extraction (see Weizman 2011; Dunlap 2018d). Bringing the point about the interpenetrated nature of these extractive logics further, the interspecies violence involved in industrial livestock mirrors—or mutually reinforces—the multiple modalities of violence necessary for controlling people and natures explored in Chap. 4. As violence unfolds increasingly in confinement—in 'concentrated animal feeding operations': concentration camps—it leaves natural environments 'devitalized', as Weis (2018) puts it, in uneasy emptiness haunted by the 'ghosts' of the animals that used to live there.

Springer (2019) argues that political ecology evinces an unfortunate blind spot, as it tends to ignore or sideline the extent to which capitalist growth and dominant speciesist worldviews are generative of massive violence against non-human animals. This renders political ecology 'decidedly anthropocentric' in Springer's assessment. Drawing upon the animal liberation movement, critical animal studies and the reinvigoration of Reclus and anarchist influences (see Chap. 3), Springer suggests that the prevailing 'liberation ecology' (Peet and Watts 1996) that foregrounds class struggle should be supplanted by a 'total liberation ecology' as an 'intersectional ethos that seeks to contest all forms of inequality and domination'. We return to this notion in the conclusion of this book. For now, it is sufficient to drive home that 'the state, capitalism, and the domination of non-human animals are interrelated processes' (Springer 2019: 3), emphasizing that this approach is already implicated in total extractivism.

Yet, whereas Springer seeks to establish total liberation ecology through engagement with anarchist thought, including that of Bookchin (1996) and its 'dialectical naturalism', he seems to sidestep converging foci among ecologically oriented Marxists, such as Moore's (2015) notion of 'the double internality' where human and more-than-human natures are interwoven in the web of life. Such sidestepping of avenues for convergence cut both ways, as Moore blissfully ignores Bookchin, Reclus and the broader anarchist current. While it appears that less acknowledged 'cracks' in anthropocentrism crosscut theoretical and political stances, the making of an 'anti-speciest political ecology' (Collard 2015) may be at the horizon.

Where Chap. 4 interrogated the coercion, pacification and the social engineering of consent, we find that non-human animals are ***biologically engineered*** for maximizing profits in economies of scale: in order to be profitably killed earlier, faster, easier. 'Animal genetics have been radically altered', writes Weis (2018: 139), 'to grow, lay, and lactate faster while tolerating confinement, which ties to the design of patentable traits, the decline of genetic diversity, and the establishment of specialized subpopulations for breeding'. With factory farming, the biological engineering of animals entails that 'animals are not only lesser subjects than humans and therefore deemed worthy of complete domination, but also objects—machines of production, bred for docility unless it clashes with other desirable attributes' (Emel and Neo 2015: 5–6). No need for engineering consent: ***toleration*** is what is required for maintaining this system of 'intensive interspecies violence' with its 'total domination of animal lives and disregard for their anguish' (Weis 2018: 140). While animals are subject of biological engineering, the social engineering of consumer populations among humans has been no less real than in the case of land deals. In the US context, Nibert (2013) documents how consumer cultures compliant with the 'needs' of the growing livestock industry was engineered in the early twentieth century (and continuing unabated since). He writes about how, in the 1920s, the 'father of public relations' Edward L. Bernays worked for ham and bacon producers in fabricating public education campaigns promoting these manifestly unhealthy foods (see Bernays 2005 [1928]). We can only imagine the devious and sophisticated technologies that will be deployed to further increase human consumption of non-humans in a future of cascading socio-ecological crisis and climate change catastrophe. Moreover, we must be critical of how human biology is changing or being engineered through the consumption of biologically engineered meats, widespread industrial pollutants and the greater implications of geoengineering[4] itself—not forgetting transhumanist interventions into the body.

The Renewable-Energy Extraction Nexus

Ecological crisis, climate change and increasing energy usage are leading to various environmental programs. Despite the proliferation of ecological concerns, high-level negotiations surrounding biodiversity loss and climate change, which includes the resulting (market-based) mitigation practices, appear to be failing at redressing these issues in any meaningful way (Dunlap and Fairhead 2014; Hunsberger et al. 2017). Capitalist growth and the financial economy remain the highest priority, which becomes increasingly apparent in the way extractivism and environmental policy interact. We are witnessing the simultaneous deployment of industrial-scale extractivism and conservation initiatives that appear to be collaborating and complementing each other in methods of land control, commodifying natural resources and extracting value, meanwhile suppressing local opposition and resistance. Highlighting this phenomenon, Büscher and Davidov (2013) reveal the 'ecotourism-extraction nexus' that demonstrates how resource extraction and eco-tourism are actually co-constructed, share similar logics and retain multiple forms of collaboration. With this angle, eco-tourism can become complementary to hydrocarbon and mineral extraction, meanwhile extractivism can pave the way to eco-tourism demarcation and revenue extraction (Duffy 2015), leaving communities to negotiate conventional and green extractive industries (Büscher and Davidov 2013; Sullivan 2013b; Hill et al. 2016). Furthermore, the discursive fabrication of 'bad' and 'good' extraction remains instrumental, as mining engaged in environmental programs and 'responsible' mining can become 'sustainable mining' or support (market-based) environmental initiatives.

Mining fabricated as 'sustainable' is based on CSR and the notion of 'offsetting'. The undeniably ecologically destructive nature of large-scale mines requires them to use the logic of offsetting that is constructed in various ways (see Dunlap and Sullivan 2019). The idea, however, is the same: if you cause damage in one place you will restore and improve

[4] See Dalby (2015).

nature in another place (See Sullivan 2009; Huff and Brock 2017). As mentioned above, environmental initiatives are further legitimizing and/or rebranding the images of companies, opening new green economic markets and undermining environmentalists opposing development projects. Offsetting has tried to brand Ilmenite (Seagle 2012), Uranium (Sullivan 2013b) and lignite coal mining (Kirsch 2014; Brock and Dunlap 2018) as 'green' and ecologically sustainable. Meanwhile, company offsetting sites, aside from rebranding operations, are then seeking to commodify and make these 'restored' or 'improved' environments legible to the green economy where biodiversity and carbon storage are then integrated into financial markets (see Sullivan 2010, 2013a; Fletcher et al. 2018; Dunlap and Sullivan 2019). Thus, as Seagle (2012) pointed out, the more companies extract and degrade environments, the more their carbon and biodiversity investments will increase in value.

Offsetting can extend to 'clean' energy projects such as solar and wind energy development. While solar and wind energy were present in the environmental initiatives of the Hambach coal mine (Brock and Dunlap 2018), we also see wind and solar projects powering the infrastructure and public imagine of mining industries, notable among them is Grupo Mexico (Dunlap 2017a, 2019a). Solar and wind energy projects are themselves paraded as ecologically sustainable, environmentally friendly and an answer to climate change mitigation and energy transition. There is little, however, to suggest that industrial-scale renewable energy can lead to solving, let alone mitigating ecological crises. In fact, it spreads electrical infrastructure across the world (see Fig. 2 of Chap. 2), energizing mass consumption and life into the Worldeater. The socio-ecological costs of solar and wind energy are under-acknowledged, if not willfully ignored, to blindly expand extractive activities and economic or (delusional) 'green' growth (see Hickel and Kallis 2019). It may be argued that, while so-called renewables (or fossil fuel+) do not exhaust the resources that they use, the plantation economy does. We disagree. First, as will be discussed below, renewables require extractivism for securing raw materials and thereby risk the possibility of exhaustion via mining, negligent recycling protocols, misleading ecological 'solution' marketing and capitalist growth imperatives. Second, and like most technological interventions, people do not know the long-term impact that wind parks have on 'the atmospheric boundary layer by (a) reducing wind speeds, (b) generating blade scale turbulence in the wake of the turbines, and (c) generating shear driven turbulence due to the reduced wind speeds in the turbine wake' (Tabassum-

Abbasi et al. 2014: 281). Moreover, absorbing the sun into solar panels—the skin of the Worldeater—and not the land risks unknown site-specific ecological alterations. Theoretical modeling and now 'concrete evidence', according to Tabassum-Abbasi et al. (2014: 273), 'is emerging that large wind farms can influence local weather but are also likely to influence the climate and can bring in significant changes'. Renewable interventions might be less abrasive than mineral and hydrocarbon extraction, yet they remain co-constructed and are producing an infrastructure intervening into wind and solar patterns on a global-scale to harness and absorb vital energy. This deserves greater consideration and research to say the least. Below we will briefly discuss five criteria—raw material extraction, land control, socio-ecological impact, energy use and decommissioning—to assess the viability of solar and wind energy projects in their respective locations.

Where do those fields of solar panels and large-metal wind towers come from? Photovoltaic (PV) solar and wind energy projects require a vast amount of extraction and processes. You will find fossil fuels at every part of the supply chain in mining, manufacturing, transportation and processing of raw materials. This also includes the increasing mechanization of large-scale mining, but consider that according to some sources, 'PV solar power systems contain approximately 5.5 tons per MW of copper, while grid energy storage installations rely on between 3 tons and 4 tons per MW' (Strong 2016). Meanwhile, according to Wind Power Monthly, 3.6 tons of copper are used per megawatt in wind energy development (Smith 2014). This includes unknown amounts of various steels (depending on the solar project) for panel frames, while wind turbines use roughly 150 metric tons of steel for reinforced concrete foundations, 250 metric tons for the rotor hubs and nacelles and 500 metric tons for the tower (Smil 2016). The production of steel, as it stands, is impossible without burning metallurgical coal—or coking coal—which is a vital raw material in the processing of renewable energy infrastructure (Dıez et al. 2002). The manufacturing of PV requires hazardous materials such as hydrochloric acid, sulfuric acid, nitric acid, hydrogen fluoride, 1,1,1-tricholoroethane, and acetone (UCS 2013; Yang et al. 2017). Thin-film PV cells also contain toxic material such as gallium arsenide, copper-indium-gallium-diselenide, cadmium-telluride and coolant liquids (Yang et al. 2017). Wind turbines also use a variety of plastics, minerals and rare earth minerals in their permanent magnets, such as dysprosium, praseodymium neodymium, and terbium (Dunlap 2018e). There are numerous variations between type of wind turbines—geared turbines that do not use permanent magnets or direct-drive turbines—and solar panels utilizing different technologies. The list of minerals and processing necessary for use are exhaustive, yet it is important to recognize the extent of the complex fossil fuel and mineral supply chains, and likely accompanying human right abuses that lurk behind so-called renewable technologies (Finley-Brook and Thomas 2011; Kiezebrink et al. 2018), which require further research, public acknowledgment and discussion—to say the least. This collaboration between extractive industries, extracting minerals for renewable energy or using renewable energy for extractive operations, has led Dunlap and Brock (2019, Dunlap 2018e) to calling this 'the renewable energy-extraction nexus'. This nexus, we would add, aids the Worldeater's colonizing drive, expanding its grid while remarkably successfully enrolling people into its entrails as it builds public opinion in accordance with selective rational interest.

Once raw materials are secured, processed and assembled, or ready for onsite assembly, the next step is to acquire land with abundant solar and wind resources. Solar and wind parks require large tracts of land (Rignall 2016; Yenneti et al. 2016; Siamanta 2017; Avila 2018; Dunlap 2017a), which opens these projects to the same issues associated with land grabbing and control: dispossession carried out by public and private security forces; land contracts dependent on deception/manipulation; acquiring contracts through coercion and, overall, multipronged counterinsurgency style interventions (Dunlap 2018a, forthcoming). Different forms of 'energy dispossession' (Baka 2017) offer various and adverse royalties, and socio-economic and ecological benefits. Land deals, and the methods used to acquire land, will depend on the companies and the politico-historical context where conventional and green extractivist projects arrive. The Union of Concerned Scientists notes that there is 'less opportunity for solar projects to share land with agricultural uses' than wind turbines, which in theory is true, yet depends entirely on the company in question (Siamanta and Dunlap, forthcoming). Meanwhile, mixing wind turbines with agriculture reportedly has severe negative impacts on agriculture and livestock (see Dunlap 2019b).

When land is acquired, the project construction begins, requiring various environmental transformations and/or degradations. Solar and wind projects require the clearing of land, which necessitates, depending on this land, different levels of deforestation, habitat loss and soil compaction with the construction of roads (Yenneti et al. 2016; Dunlap 2017a, 2018a, 2019b; Yang et al. 2018). Wind turbines have concrete foundations that, depending on the geographic topology, are between 7 and 14 meters (32–45 ft.) deep and about 1 and –21 meters (52–68 ft.) in diameter, and, along with road, subterranean or above-ground power lines, these will also have significant environmental impacts (Dunlap 2017a, 2019b). Again, depending on the topology, hydrology and the level of contestation, the environmental impacts will have various intensities. While wind turbines have reportedly leaked oil (lubricating the turbines) into the ground and open wells (Dunlap 2019b), solar plants risk leaking coolant liquids (Yang et al. 2018). Where animals graze this has led to numerous reports from locals claiming that oil leaking from wind turbines has impacted the health, reproduction and has even killed cattle (see Dunlap 2019b). Wind parks are known for killing avian species—birds and bats (Tabassum-Abbasi et al. 2014)—and solar projects contribute to four types of light pollution: urban sky glow, light trespass, glare and clutter (see Yang et al. 2017). While hydrocarbon power plants have numerous operating and extractive costs, renewable energy also has various ecological impacts, the severity of which depends on geographic location, park density and mitigation measures.

The social impacts are similar to any large-scale development project. Rural areas are going to experience an influx of foreigners, habits and money, including project managers, skilled laborers and temporary or permanent jobs. This includes social development funds, infrastructure projects and various individual and collective benefits depending on the methods of land acquisition. Development projects, in turn, attract increases in crime, drug-use and prostitution and, overall, forms of rural gentrification (Dunlap 2017a), which has distinct similarities with mining projects (see Kirsch 2014: 31). This is compounded by the proximity of solar, wind or other extractive operations to towns, people and collectively used resources such as water (Yenneti et al. 2016; Dunlap 2019b). These factors are key to determining the severity of health issues, which requires further research.

Next, what is this energy used for? Is this energy providing light and heat to homes or is it fueling socially and ecologically destructive industries? The Isthmus of Tehuantpec region of Oaxaca, Mexico—known for having some of the best wind energy resources in the world—we find powering Wal-Mart, CENMEX, Grupo Bimbo (low-grade food), mining companies and various industrial construction companies. Additionally, this includes exporting energy to Guatemala, Belize and the United States, while towns are engulfed by wind turbines and electricity prices are increasing (Dunlap 2019b). There are various mining and construction companies as well as Google who are utilizing solar, wind and other renewable energy sources for their operations. But it is our contention that socio-ecological destruction will prevail unless the legally binding profit-maximization imperative of corporations is eliminated and rolled back, or economic growth is re-conceptualized and industrial societies actively implement genuine policies of degrowth (see D'Alisa et al. 2014; Kallis 2018). Harmonious socio-ecological relationships do not need more Wal-Marts, plastic trinkets and mines than the years before; ecological catastrophe suggests otherwise and remains indications of the Worldeater. If energy transition is at best about economizing exploitation and at worst expanding resource frontiers, then renewable energy is in fact 'Fossil Fuel+' (Dunlap 2018c): renewing capitalism, resource colonization and ecological destruction.

Finally, decommissioning solar and wind energy projects after 25–35 years: the high-grade industrial toxics associated with photovoltaic solar leads to various concerns about proper disposal and recycling methods, which companies are hesitant to take up (Aman et al. 2015). The EU imposed a mandatory producer recycling rules in 2012, which led companies to exit European Markets. Among these companies was First Solar—now the operator of the California Flats Project that powers Apple headquarters complex among other things—estimated to generate enormous amounts of toxic waste in 20–30 years to come (Huffington Post 2015). Both solar and wind turbines have a 25–35-year shelf life, raising serious concerns about reusing and recycling copper, steel, plastic, concrete and magnets (Guezuraga et al. 2012; Hoenderdaal et al. 2013), which requires detailed, holistic and ethnographically based life-cycle assessments as opposed to models over-emphasizing carbon at the expense of other industrial wastes. Every single digging machine, processing chemical and component in a solar panel and wind turbine are going to have an ecologically (degrading) 'social life' along global production chains, which is narrowed by only focusing on carbon dioxide accounting (see Dunlap 2018e). Decommissioning, recycling and ecologically responsibly disposing of waste are central issues, yet we need to ask what type of world solar, wind and other industrial-scale technologies are actually renewing, if not creating.

Conclusion

This chapter has traced the Worldeater in process: its movements, expansions and extractive practices as it spreads and accumulates its 'body'—infrastructural grid—across the surface of, and into, the earth. This has been an attempt at revealing the imperative of total extractivism at the current conjuncture of the techno-capitalist world system. While such a description necessarily fails in grasping the entirety of the total character of these processes, we have interrogated a wide range of issues as we have traced the Worldeater as it incorporates rural people—entrails making—into its massively expanding body of mines and plantations. Its entrails, we have shown, are already filled to the brim with corpses of sentient beings, but its appetite for non-human animals show few signs of subsiding as factory farm infrastructure covers larger and larger parts of the world.

The imperative of total extractivism, moreover, reveals the false claims of 'greening' as the renewable energy-extraction nexus suggests. The green economy emerges as a worldeating device and a violent technology of extraction. Few fabrications are more successful at present than the pretentions of this green economy, pretentions that serve the Worldeater's interests as they help devising what Taussig (1980: 28) described as consciousness subjugated by the commodity form. This in turn contributes to enabling the possession of people who then internalize and re-project the operations of the Worldeater and its colonizing—even terraforming—machinery without question. Consciousness is subjugated, domination by commodities is normalized. Un-subjugated consciousness is left isolated and fragmented, yet retains the ability of sabotage, 'monkey wrenching' and violently tinkering with the Worldeater's machinery. Cunning and conscious miscalculations—individual and collective actions—retain the possibility to disrupt, delay and push back what is in fact destroying people—often by means of excess and enchantment—and the earth. We now turn to conclude the book with reflections on human power in the face of the Worldeater, outlining in a more hopeful mode some of the possible ways out of the entrails.

Bibliography

Acosta, A. (2013 [2011]). Extractivism and Neoextractivism: Two Sides of the Same Curse. In M. Lang & D. Mokrani (Eds.), *Beyond Development: Alternative Visions from Latin America* (pp. 61–86). Amsterdam: Transnational Institute.

Aman, M., Solangi, K., Hossain, M., et al. (2015). A Review of Safety, Health and Environmental (SHE) Issues of Solar Energy System. *Renewable and Sustainable Energy Reviews, 41*, 1190–1204.

Avila, S. (2018). Environmental Justice and the Expanding Geography of Wind Power Conflicts. *Sustainability Science 13*(3), 599–616.

Baka, J. (2017). Making Space for Energy: Wasteland Development, Enclosures, and Energy Dispossessions. *Antipode, 49*, 977–996.

Bebbington, A. (2012). *Social Conflict, Economic Development and the Extractive Industry: Evidence from South America*. London: Routledge.

Bernays, E. (2005 [1928]). *Propaganda, with an Introduction by Michal Crispin*. New York: Ig Publishing.

Bernstein, H. (2010). *Class Dynamics of Agrarian Change*. London: Kumarian Press.

Birss, M. (2017). Criminalizing Environmental Activism: As Threats to the Environment Increase Across Latin America, New Laws and Police Practices Take Aim Against the Front Line Activists Defending Their Land and Resources. *NACLA Report on the Americas, 49*(3), 315–322.

Bonneuil, C., & Fressoz, J.-B. (2016). *The Shock of the Anthropocene: The Earth, History and Us*. New York: Verso Books.

Bookchin, M. (1996). *The Philosophy of Social Ecology: Essays on Dialectical Naturalism*. Montreal: Black Rose Books.

Borras, S. M., & Franco, J. (2013). Global Land Grabbing and Political Reactions 'From Below'. *Third World Quarterly, 34*, 1723–1747.

Brock, A., & Dunlap, A. (2018). Normalising Corporate Counterinsurgency: Engineering Consent, Managing Resistance and Greening Destruction Around the Hambach Coal Mine and Beyond. *Political Geography, 62*, 33–47.

Büscher, B., & Davidov, V. (2013). *The Ecotourism-Extraction Nexus: Political Economies and Rural Realities of (Un)Comfortable Bedfellows*. London: Routledge.

Chigumira, E. (2018). Political Ecology of Agrarian Transformation: The Nexus of Mining and Agriculture in Sanyati District, Zimbabwe. *Journal of Rural Studies, 61*, 265–276.

Collard, R.-C. (2015). Ethics in the Human. In T. Perreault et al. (Eds.), *The Routledge Handbook of Political Ecology* (pp. 127–139). New York: Routledge.

Crosby, A. C., & Monaghan, J. (2018). *Policing Indigenous Movements: Dissent and the Security State*. Black Point: Fernwood Publishing.

D'Alisa, G., Demaria, F., & Kallis, G. (2014). *Degrowth: A Vocabulary for a New Era*. London: Routledge.

Dalby, S. (2015). Geoengineering: The Next Era of Geopolitics? *Geography Compass, 9*, 190–201.

Demaria, F., & Kothari, A. (2017). The Post-Development Dictionary Agenda: Paths to the Pluriverse. *Third World Quarterly, 38*, 2588–2599.

Dıez, M., Alvarez, R., & Barriocanal, C. (2002). Coal for Metallurgical Coke Production: Predictions of Coke Quality and Future Requirements for Cokemaking. *International Journal of Coal Geology, 50*, 389–412.

Duffy, R. (2015). Nature-Based Tourism and Neoliberalism: Concealing Contradictions. *Tourism Geographies, 17*, 529–543.

Dunlap, A. (2017a). 'The Town Is Surrounded': From Climate Concerns to Life Under Wind Turbines in La Ventosa, Mexico. *Human Geography, 10*, 16–36.

Dunlap, A. (2018a). Counterinsurgency for Wind Energy: The Bíi Hioxo Wind Park in Juchitán, Mexico. *The Journal of Peasant Studies, 45*, 630–652.

Dunlap, A. (2018b). Insurrection for Land, Sea and Dignity: Resistance and Autonomy Against Wind Energy in Álvaro Obregón, Mexico. *Journal of Political Ecology, 25*, 120–143.

Dunlap, A. (2018c). The 'Solution' Is Now the 'Problem': Wind Energy, Colonization and the 'Genocide-Ecocide Nexus' in the Isthmus of Tehuantepec, Oaxaca. *The International Journal of Human Rights, 42*, 550–573.

Dunlap, A. (2018d). "A Bureaucratic Trap": Free, Prior and Informed Consent (FPIC) and Wind Energy Development in Juchitán, Mexico. *Capitalism Nature Socialism, 29*, 88–108.

Dunlap, A. (2018e). *Reconsidering the Logistics of Autonomy: Ecological Autonomy, Self-Defense and the Polícia Comunitaria in Álvaro Obregón, Mexico.* Retrieved from https://www.tni.org/files/article-downloads/erpi_cp_8_dunlap.pdf.

Dunlap, A. (2019a). 'Agro sí, mina NO!' The Tía Maria Copper Mine, State Terrorism and Social War by Every Means in the Tambo Valley, Peru. *Political Geography, 71*, 10–25.

Dunlap, A. (2019b). *Renewing Destruction: Wind Energy Development in Oaxaca, Mexico.* London: Rowman & Littlefield.

Dunlap, A. (forthcoming). Wind, Coal and Copper: The Politics of Land Grabbing, Counterinsurgency and the Social Engineering of Extraction. *Globalizations.*

Dunlap, A., & Brock, A. (2019). When the Wolf Guards the Sheep: Green Extractivism in Germany and Mexico. In S. Springer, M. Locret, J. Mateer, et al. (Eds.), *Anarchist Political Ecology* (Vol. 3). Oakland, CA: PM Press.

Dunlap, A., & Fairhead, J. (2014). The Militarisation and Marketisation of Nature: An Alternative Lens to 'Climate-Conflict'. *Geopolitics, 19*, 937–961.

Dunlap, A., & Sullivan, S. (2019). A Faultline in Neoliberal Environmental Governance Scholarship? Or, Why Accumulation-by-Alienation Matters. *Environment & Planning E: Nature and Space*, 1–19.

Emel, J., & Eds, H. N. (2015). *Political Ecologies of Meat.* London: Routledge.

Escobar, A. (2012 [1995]). *Encountering Development: The Making and Unmaking of the Third World.* Princeton, NJ: Princeton University Press.

Esteva, G. (2014). Commoning in the New Society. *Community Development Journal, 49*, 144–159.

Fairbairn, M., Fox, J., Isakson, S. R., et al. (2014). Introduction: New Directions in Agrarian Political Economy. *Journal of Peasant Studies, 41*, 653–666.

Finley-Brook, M., & Thomas, C. (2011). Renewable Energy and Human Rights Violations: Illustrative Cases from Indigenous Territories in Panama. *Annals of the Association of American Geographers, 101*, 863–872.

Fletcher, R., Dressler, W. H., Anderson, Z. R., et al. (2018). Natural Capital Must Be Defended: Green Growth as Neoliberal Biopolitics. *The Journal of Peasant Studies, 46*, 1–28.

Gamu, J. K., & Dauvergne, P. (2018). The Slow Violence of Corporate Social Responsibility: The Case of Mining in Peru. *Third World Quarterly, 39*, 959–975.

Gerber, J.-F. (2011). Conflicts Over Industrial Tree Plantations in the South: Who, How and Why? *Global Environmental Change, 21*, 165–176.

Giarracca, N., & Teubal, M. (2014). Argentina: Extractivist Dynamics of Soy Production and Open-Pit Mining. In H. Veltmeyer & J. F. Petras (Eds.), *The New Extractivism: A Post-Neoliberal Development Model or Imperialism of the Twenty-First Century?* (pp. 47–79). London: Zed Books.

González-Hidalgo, M., & Zografos, C. (2017). How Sovereignty Claims and "Negative" Emotions Influence the Process of Subject-Making: Evidence from a Case of Conflict Over Tree Plantations from Southern Chile. *Geoforum, 78*, 61–73.

Gudynas, E. (2009). Diez tesis urgentes sobre el nuevo extractivismo. In J. Schuldt, A. Acosta, A. Barandiarán, et al. (Eds.), *Extractivismo, política y sociedad*. Quito: Centro Andino de Acción Popular (CAAP) and Centro Latinoamericano de Ecología Social (CLAES).

Gudynas, E. (2010). The New Extractivism of the 21st Century: Ten Urgent Theses About Extractivism in Relation to Current South American Progressivism. *Americas Program Report, 21*, 1–14.

Gudynas, E. (2013 [2011]). Debates on Development and Its Alternatives in Latin America: A Brief Heterodox Guide. In M. Lang & D. Mokrani (Eds.), *Beyond Development: Alternative Visions from Latin America* (pp. 15–40). Amsterdam: Transnational Institute.

Guezuraga, B., Zauner, R., & Pölz, W. (2012). Life Cycle Assessment of Two Different 2 MW Class Wind Turbines. *Renewable Energy, 37*, 37–44.

Gunderson, R. (2011). From Cattle to Capital: Exchange Value, Animal Commodification, and Barbarism. *Critical Sociology, 39*, 259–275.

Hall, R., Edelman, M., Borras, S. M., Jr., Scoones, I., White, B., & Wolford, W. (2015). Resistance, Acquiescence or Incorporation? An Introduction to Land Grabbing and Political Reactions 'From Below'. *Journal of Peasant Studies, 42*(3-4), 467–488.

Hall, D., Hirsch, P., & Li, T. M. (2011). *Powers of Exclusion: Land Dilemmas in Southeast Asia*. Honolulu: University of Hawaii Press.

Haraway, D. (2015). Anthropocene, Capitalocene, Plantationocene, Chthulucene: Making Kin. *Environmental Humanities, 6*(1), 159–165.

Harvey, D. (2003). *The New Imperialism*. New York: Oxford University Press.

Hickel, J. L., & Kallis, G. (2019). Is Green Growth Possible? *New Political Economy*. https://doi.org/10.1080/13563467.2019.1598964.

Hill, W., Byrne, J., & de Vasconcellos Pegas, F. (2016). The Ecotourism-Extraction Nexus and Its Implications for the Long-Term Sustainability of Protected Areas: What Is Being Sustained and Who Decides? *Journal of Political Ecology, 23*, 308–327.

Hoenderdaal, S., Espinoza, L. T., Marscheider-Weidemann, F., et al. (2013). Can a Dysprosium Shortage Threaten Green Energy Technologies? *Energy, 49*, 344–355.

Hribal, J. (2010). *Fear of the Animal Planet: The Hidden History of Animal Resistance*. Petrolia and Oakland, CA: CounterPunch and AK Press.

Huff, A., & Brock, A. (2017). *Intervention—"Accumulation by Restoration: Degradation Neutrality and the Faustian Bargain of Conservation Finance"*. Retrieved from https://antipodefoundation.org/2017/11/06/accumulation-by-restoration/.

Hunsberger, C., Corbera, E., Borras, S. M., Jr., et al. (2017). Climate Change Mitigation, Land Grabbing and Conflict: Towards a Landscape-Based and Collaborative Action Research Agenda. *Canadian Journal of Development Studies/Revue canadienne d'études du développement 38*(3), 305–324.

Illich, I. (1970). Planned Poverty: The End Result of Technical Assistance. In *Celebration of Awareness: A Call for Institutional Revolution*. London: Marion Boyars.

Jakobsen, J. (2018). Towards a Gramscian Food Regime Analysis of India's Agrarian Crisis: Counter-Movements, Petrofarming and Cheap Nature. *Geoforum, 90*, 1–10.

Jakobsen, J., & Hansen, A. (2019). Geographies of Meatification: An Emerging Asian Meat Complex. *Globalizations*, 1–17. https://doi.org/10.1080/14747731.2019.1614723.

Kallis, G. (2018). *Degrowth*. New York: Columbia University Press.

Kiezebrink, V., Wilde-Ramsing, J., & Kate Gt. (2018). *Human Rights in Wind Turbine Supply Chains: Towards a Truly Sustainable Energy Transition*. Retrieved from https://www.somo.nl/wp-content/uploads/2018/01/Final-ActionAid_Report-Human-Rights-in-Wind-Turbine-Supply-Chains.pdf.

Kirsch, S. (2014). *Mining Capitalism: The Relationship Between Corporations and Their Critics*. Berkeley, CA: University of California Press.

Kröger, M. (2016). The Political Economy of 'Flex Trees': A Preliminary Analysis. *The Journal of Peasant Studies, 43*(4), 886–909.

Lahiri-Dutt, K. (2018a). Extractive Peasants: Reframing Informal Artisanal and Small-Scale Mining Debates. *Third World Quarterly 39*(8), 1561–1582.

Lahiri-Dutt, K. (Ed.). (2018b). *Between the Plough and the Pick: Informal Mining in the Contemporary World*. Canberra: Australian National University Press.

Li, T. M. (2014). *Land's End: Capitalist Relations on an Indigenous Frontier*. Durham, NC: Duke University Press.

Li, T. M. (2018). After the Land Grab: Infrastructural Violence and the "Mafia System" in Indonesia's Oil Palm Plantation Zones. *Geoforum, 96*, 328–337.

Malm, A. (2016). *Fossil Capital: The Rise of Steam Power and the Roots of Global Warming*. London: Verso Books.

Marston, A., & Perreault, T. (2017). Consent, Coercion and Cooperativismo: Mining Cooperatives and Resource Regimes in Bolivia. *Environment and Planning A: Economy and Space, 49*(2), 252–272.

McKay, B. M. (2017). Agrarian Extractivism in Bolivia. *World Development, 97*, 199–211.

McKittrick, K. (2011). On Plantations, Prisons, and a Black Sense of Place. *Social & Cultural Geography, 12*(8), 947–963.

McMichael, P. (2013). *Food Regimes and Agrarian Questions*. Rugby: Fernwood Publishing and Practical Action Publishing.

Middeldorp, N., Morales, C., & van der Haar, G. (2016). Social Mobilisation and Violence at the Mining Frontier: The Case of Honduras. *The Extractive Industries and Society, 3*, 930–938.

Mintz, S. (1985). *Sweetness and Power*. New York: Viking Penguin.

Moore, J. W. (2010). 'Amsterdam Is Standing on Norway' Part I. *Journal of Agrarian Change, 10*(1), 33–68.

Moore, J. W. (2015). *Capitalism in the Web of Life: Ecology and the Accumulation of Capital*. London: Verso.

Nibert, D. A. (2013). *Animal Oppression and Animal Violence: Domesecration, Capitalism, and Global Conflict*. New York: Columbia University Press.

Noske, B. (1989). *Humans and Other Animals: Beyond the Boundaries of Anthropology*. London: Pluto Press.

Peet R and Watts M. Eds. (1996) *Liberation Ecologies: Environment, Development, Social Movements*. London and New York: Routledge.

Peluso, N. L. (2017). Plantations and Mines: Resource Frontiers and the Politics of the Smallholder Slot. *The Journal of Peasant Studies, 44*, 834–869.

Perreault, T. (2018). The Plantation and the Mine: Comment on "After the Land Grab: Infrastructural Violence and the 'Mafia System' in Indonesia's Oil Palm Plantation Zone" by Tania Li. *Geoforum, 96*, 345–347.

Petras, J., & Veltmeyer, H. (2014). *Extractive Imperialism in the Americas*. London: Brill.

Post, H. (2015). *The Dark Side of the Sun: Solar Power and Global Electronic Waste*. Retrieved from https://www.huffingtonpost.com/peter-schwartz/solar-power-and-the-looming_b_7104058.html.

Rahnema, M., & Bawtree, V. (1997). *The Post-Development Reader*. London: Zed Books.

Rignall, K. E. (2016). Solar Power, State Power, and the Politics of Energy Transition in Pre-Saharan Morocco. *Environment and Planning A, 48*(3), 540–557.

Rodgers, D., & O'Neill, B. (2012). Infrastructural Violence: Introduction to the Special Issue. *Ethnography, 13*, 401–412.

Rosales, A. (2016). Deepening Extractivism and Rentierism: China's Role in Venezuela's Bolivarian Developmental Model. *Canadian Journal of Development Studies/Revue canadienne d'études du développement, 37*(4), 560–577.

Seagle, C. (2012). Inverting the Impacts: Mining, Conservation and Sustainability Claims Near the Rio Tinto/Qmm Ilmenite Mine in Southeast Madagascar. *Journal of Peasant Studies, 39*, 447–477.

Siamanta, Z. C. (2017). Building a Green Economy of Low Carbon: The Greek Post-Crisis Experience of Photovoltaics and Financial 'Green Grabbing'. *Journal of Political Ecology, 24*, 258–276.

Siamanta, C., & Dunlap, A. (forthcoming). *'Accumulation by Wind Energy': Wind Energy Development as a Capitalist Trojan Horse in Crete*, Greece and Oaxaca, Mexico, ACME

Smil, V. (2016). *To Get Wind Power You Need Oil.* Retrieved from http://spectrum.ieee.org/energy/renewables/to-get-wind-power-you-need-oil.

Smith, P. (2014). *Soaring Copper Prices Drive Wind Farm Crime.* Retrieved from http://www.windpowermonthly.com/article/1281864/soaring-copper-prices-drive-wind-farm-crime.

Springer, S., Locret, M., Mateer, J., et al. (2019). *Anarchist Political Ecology.* Oakland, CA: PM Press.

Strong, Z. (2016). *Will the Transition to Renewable Energy Be Paved in Copper?* Retrieved from https://www.renewableenergyworld.com/articles/2016/01/will-the-transition-to-renewable-energy-be-paved-in-copper.html.

Stuart, D., Schewe, R. L., & Gunderson, R. (2013). Alienation in the Dairy Sector. *Sociologia Ruralis, 53*, 201–222.

Sullivan, S. (2009). Green Capitalism, and the Cultural Poverty of Constructing Nature as Service Provider. *Radical Anthropology, 3*, 18–27.

Sullivan, S. (2010). 'Ecosystem Service Commodities'—A New Imperial Ecology? Implications for Animist Immanent Ecologies, with Deleuze and Guattari. *New Formations: A Journal of Culture/Theory/Politics, 69*, 111–128.

Sullivan, S. (2013a). Banking Nature? The Spectacular Financialisation of Environmental Conservation. *Antipode, 45*, 198–217.

Sullivan, S. (2013b). After the Green Rush? Biodiversity Offsets, Uranium Power and the 'Calculus of Casualties' in Greening Growth. *Human Geography, 6*, 80–101.

Svampa, M. (2013 [2011]). Resource Extractivism and Alternatives: Latin American Perspectives on Development. In M. Lang & D. Mokrani (Eds.), *Beyond Development: Alternative Visions from Latin America* (pp. 117–144). Amsterdam: Transnational Institute.

Tabassum-Abbasi, P. M., Abbasi, T., et al. (2014). Wind Energy: Increasing Deployment, Rising Environmental Concerns. *Renewable and Sustainable Energy Reviews, 31*, 270–288.

Taussig, M. T. (1980). *The Devil and Commodity Fetishism in South America.* Durham, NC: University of North Carolina Press.

Torres, B. (2007). *Making a Killing: The Political Economy of Animal Rights*. Oakland: AK Press.

Twine, R. (2012). Revealing the 'Animal-Industrial Complex'—A Concept & Method for Critical Animal Studies? *Journal for Critical Animal Studies, 10*(1), 12–39.

UCS. (2013). Environmental Impacts of Solar Power. Retrieved from https://www.ucsusa.org/clean_energy/our-energy-choices/renewable-energy/environmentalimpacts-solar-power.html#.W-NS9TFMHVk.

Vergara-Camus, L., & Kay, C. (2018). New Agrarian Democracies: The Pink Tide's Lost Opportunity. *Socialist Register, 54*, 224–243.

Wadiwel, D. (2015). *The War Against Animals*. Leiden: Brill.

Walker, R. A. (2004). *The Conquest of Bread: 150 Years of Agribusiness in California*. New York: The New Press.

Watts, M. (2004). Resource Curse? Governmentality, Oil and Power in the Niger Delta, Nigeria. *Geopolitics, 9*(1), 50–80.

Weis, T. (2010). The Accelerating Biophysical Contradictions of Industrial Capitalist Agriculture. *Journal of Agrarian Change, 10*, 315–341.

Weis, T. (2013). *The Ecological Hoofprint: The Global Burden of Industrial Livestock*. London: Zed Books.

Weis, T. (2016). Toward 120 Billion: Dietary Change and Animal Lives. *Radical Philosophy, 199*, 8–13.

Weis, T. (2018). Ghosts and Things: Agriculture and Animal Life. *Global Environmental Politics, 18*(2), 134–142.

Weizman, E. (2011). *The Least of All Possible Evils: Humanitarian Violence from Arendt to Gaza*. London: Verso.

White, R. J. (2017). Capitalism and the Commodification of Animals: The Need for Critical Vegan Praxis, Animated by Anarchism! In D. Nibert (Ed.), *Animal Oppression and Capitalism*. Connecticut: Praeger.

White, B., Borras, S. M., Hall, R., et al. (2012). The New Enclosures: Critical Perspectives on Corporate Land Deals. *Journal of Peasant Studies, 39*, 619–647.

Xu, Y. (2019). Politics of Inclusion and Exclusion in the Chinese Industrial Tree Plantation Sector: The Global Resource Rush Seen from Inside China. *The Journal of Peasant Studies 46*(4), 767–791.

Yang, H.-J., Lim, S.-Y., & Yoo, S.-H. (2017). The Environmental Costs of Photovoltaic Power Plants in South Korea: A Choice Experiment Study. *Sustainability, 9*, 1773.

Ye, J., van der Ploeg, J. D., Schneider, S., et al. (2019). The Incursions of Extractivism: Moving from Dispersed Places to Global Capitalism. *The Journal of Peasant Studies*, 1–29.

Yenneti, K., Day, R., & Golubchikov, O. (2016). Spatial Justice and the Land Politics of Renewables: Dispossessing Vulnerable Communities Through Solar Energy Mega-Projects. *Geoforum, 76*, 90–99.

CHAPTER 6

Conclusion: Out of the Entrails—Reflections on Human Power

The soil will revolt when stressed through monoculture, and as livelihoods are threatened from poor crop yields, so too will the people.
—Simon Springer

If you love, you must fight.
—Ward Churchill

Abstract The book concludes with reflections on human power in the face of the Worldeater, outlining in a more hopeful mode some of the possible ways out of its entrails. The conclusion responds to the crucial question: do humans have the power to resist the allure of the Worldeater and escape its entrails—and what can they do? The conclusion advocates the idea of total liberation in response to total extractivism. This chapter emphasizes the need for looking critically at the everyday normalization of hierarchies and oppression in sustaining the techno-capitalist trajectory. Suggesting avenues for taking steps toward disrupting and decomposing the Worldeater(s) parts, the conclusion draws on insights from green anarchism, degrowth and bioregionalism, not as ideals for emulation but as sources of inspiration and strength for people to act, come together and struggle against the Worldeater(s).

Keywords Resistance • Capitalism • Anarchism • Total liberation • Degrowth

A. Dunlap, J. Jakobsen, *The Violent Technologies of Extraction*,
https://doi.org/10.1007/978-3-030-26852-7_6

Chapter 2 began by flirting with the reality of worms, octopuses and the Leviathanic beast to discuss socio-ecological catastrophe at the hands of the Worldeater. Framing the book this way allowed us to put forward a mythical conception to describe the global capitalist imperative—or permanent tension—toward total extractivism. The Worldeater concept, we hope, opens the door to a pluriverse of perspectives, transcending the positivist and materialist discourses of modernity and Marxian thought. We see these traditions as complementary, even if imbalanced and limited, where we hope the Worldereater framing can affirm existing spaces, open new doors and strengthen 'the post-development convergence' (see Nirmal and Rocheleau 2019). The Worldeater situated our review and discussion centered on critical agrarian studies and political ecology in Chap. 3 that we felt are the interdisciplinary studies that share the most affinity and aptitude for coming to intellectual daggers with the Worldeater. Again, we hope that these studies will take up the pluriversal challenge of considering the psycho-social, emotional and—even—spiritual power associated with techno-capitalist progress that the concept of the Worldeater tries to convey. Situating, if not clearing some disciplinary brush, Chap. 4 allowed us to examine the latest 'chewing' and 'claw swipes' of the Worldeater in its demarcation of enclosures and extracting resources with coercive and enchanting forces—the arts of scientific violence, counterinsurgency. This involves the political reactions 'from above' as it relates to the conventional and so-called green military, police and extra-judicial forces, but also 'soft' pacification techniques. Matching this conventional and 'green' articulation, Chap. 5 explored and identified the forming nexus between conventional and renewable natural resource extraction that is emblematic of the global capitalist economy's drive toward total resource extractivism—worldeating. Here we chart the different formations, subjectifications and articulations of the renewable-energy extraction nexus, which we hope encourages greater exploration and inquiry.

The situation humanity faces—in all of its billions of unique multiplicities—is not all that different from what Hansel and Gretel confronted. When they were separated from their families, a creepy person in a house full of candy invited them in and took care of them, letting them eat the candy and junk food on display around the house, but also feeding them chicken and other three course meals. As the story goes, this sweet allure and kindness was really just to satisfy the desire for roasted children. In the Hansel and Gretel nursery rhyme, the kids flip the script and kick the 'creepy person'—a story engaged in witch bashing[1]—into the oven, cook them and run away. This creepy person, we might consider, embodies the ethos of the Worldeater. If humans and non-humans are not being expelled, then they are being fattened up in the factory farms, or enchanted to be cooked or, more accurately, kept alive just long enough—using their rational best interests—to continue a particular type of work, managerial role, lifestyle and, overall, participation in the techno-industrial system. Maybe this is following migration routes to find work or people trying not

[1] This old creepy woman was depicted as a witch and falls into Judeo-Christian genocidal strategies of erasing land-based knowledges/hermetic sciences and the suppression of—woman or non-gendered—people (and all their powers).

to kill themselves in banal—or 'bullshit' (Graeber 2018)—office jobs and so on. The variation and possibility of lived situations are enormous. The point, however, is that humans have broken ecological balance and are systematically poisoning their habitats and this is done by betraying their traditional tree, animal and river friends—to name only a few—in exchange for modernity and economic growth or 'second nature' (O'Connor 1988; Escobar 1996). The story of Hansel and Gretel—in keeping with the playful yet (deadly) serious spirit invoked throughout this book—in our reality may not be the happy ending told to children, as in a similar story mainstream society reproduces with notions of eco-modernism and green growth. This raises the question: how will people get out of the candy house—how will they get out of the temple of techno-capitalist progress that has been constructed over the last three centuries, if not longer? Said differently, do humans have the power to resist the allure of the Worldeater and escape its entrails—what can they do?

In theory the answer is, 'yes'. In practice the future is undetermined, but if the past century is any indication it is, 'no'. While this book has largely served as a negative treatment of the current state of the world and the research guiding it, the purpose, however, was twofold. First, we have sought to reconceive techno-capitalist progress in the negative terms as the Worldeater(s) to honor pre-capitalist mythology and stress the past, present and future trajectory that industrial development has created. This we contend challenges the category and ontology of the disputed term Anthropocene. The Worldeater(s), while open ended in its origin, specifically locates the cultural values and technologies responsible for climate and ecological catastrophe or *Worldeating*, thus avoiding human-centrism and homogenizing human participation in constructing the Worldeater(s) yet pushing toward the urgency of subversion against techno-capitalist progress. Second, and more hopefully, we have been striving to illuminate and/or propose key features that should be subverted, resisted and undermined to fight possession by the spirit of the Worldeater.

Ellul and the anarchists or, more specifically, green anarchists are right. If humans are to have the strength to confront the ethos of the Worldeater in all of its manifestations, they must take seriously the organization of colonial/statist society. Specifically, this requires maintaining permanent tensions against human (racism, sexism, all-phobia) and non-human hierarchies (speciesism/human supremacy), symbolic culture, patriarchy and divisions of labor/specialization. All of these technologies in the widest sense—and in all of their variants—create allure and enchantment. People enjoy what these technologies—and by implication capitalism—offers. The first reaction against this book will likely be about the undeniable benefits of western medicine (biomedicine), poverty reduction statistics, airplanes, computers and overall enchantment with technological progress. While the benefits (to some degree) are undeniable, they are overestimated and do not take into account their incalculable costs associated with techno-industrial development. 'Modern mechanical science', as Merchant (1983) and Shiva (2002 [1989]) accurately call it, has displaced and discredited centuries old medicinal practices and Indigenous traditions with their extensive knowledge-base of the natural world, its innumerable variety of species and their interrelations. Examples of the domination of scientism could easily be multiplied, but the whole business of 'estimating' cost-benefits on these matters is largely beside the point. What we want to drive home, however, is that the green anarchist stance and praxis of permanent tensions against the enchantment with the techno-industrial world should be taken seriously. Only through such tensions—embodying defiance and refusal to let oneself proceed along the prevailing order of things—can consciousness re-emerge from its submergence in the Beast's entrails.

There are endless technological allures, and anthropological navel gazing becomes useful to see how addiction and dependence to these social constructions blend and cross between categories to exist and be carried forward by people. That said, the techno-linear vision might be at the root of human separation with non-human nature (Dunlap 2014b; Romanyshyn 1989), which means this vision should be changed, possibly even regarded as a psycho-social pathology even if presently applauded by the institutional regimes in place. And again, the point is that while these technologies of development have been enforced and imposed to create a structure of conquest, we can say that the Worldeater or techno-capitalist progress has equally been shaped and supported by allure and enchantment—fixations with big machines, bridges and infrastructures of convenience in general as examples of human accomplishment. At risk of being reductionist, this accomplishment, however, we can say exerts control over environments, whether it is dams controlling rivers, leveling trees to promote ecosystems legibility, building skyscrapers and so on. This is not to detract unjustly from the more practical rationales for these activities, but to recognize the affinity between political control and extraction. The proposal is to look critically at all these things everywhere and to take incremental steps to crease the socio-ecologically degradation associated with technological and infrastructural development. Do not take this as advocating that people become self-deputized (and righteous) political police to their friends and family, but to find ways to undermine—individually or collectively—these technologies and the existence of all police in the first place with the intention to banish cybernetic bureaucracies and restore peoples relationships with their habitats—all the trees, plants and animals—and each other.

All these labels aside, we must learn to care for everyone—especially those taking genuine positions to stop socio-ecological catastrophe.[2] This 'everyone' really intends to include the animals, insects, rivers and neglected people—human and non-human—more than to express care for CEOs and earth destroyers. As Derrick Jensen (2006)—despite all his faults (see ISIW 2014; Matisons and Ross 2015)—reminded us, 'you cannot argue with sociopaths, you can't argue with fascists and you can't argue with those who are benefitting from an economic system—you have to stop them through some form of force, and that force can be violent or non-violent'[3]. This is to say, care has its limitations and can be directed in various ways.

This expansive care also coincides with learning to respect the land and each other, but this might not include private property and public spaces. Moreover, we must learn from and respect children—the Hansel and Gretels of this world and their innumerable siblings—whose imaginations are still not submerged and turned into entrails. We must learn to care and nourish the vitality that surrounds us and stop killing and replacing this vitality with dead things or dead things that need electricity—from other processes of mining, killing and capturing—to run through them to make matter vital. We might consider electricity as the resurrection of the dead. Whether this is fossilized plants, animals, humans and dinosaurs or even people's labor power or the body count related to work stress and industrial accidents (see House and Square 2002; Glendinning 1990), electricity in its dominate (industrial) mode of production—and the Worldeater in general—is fueled by death.

To clarify, this is not about discarding, but recognizing, the power of 'things' and 'objects'. Of course it is possible to love and cherish dead matter, even becoming en$laved by it—this is not a denial that things matter; on the contrary, it is that they matter too much and are colonizing

[2] Prevailing structures of power do the opposite: land and environmental defenders as well as individuals seeking to stop the trajectory of ecological destruction through direct action are routinely imprisoned or killed across the world.

[3] See minute 2:40, available at: https://www.youtube.com/watch?v=ssYBZmK9hmA.

and consuming people and the world. Things inhibit and condition the agency of the living. There does not need to be a total rejection of technology and commodities, but there needs to be a serious and careful rebalancing of priorities, relationships, 'tastes' and engagement with the world around us—that is, if we want to stop it from burning and its inhabitants turning inward and eating themselves both metaphorically and literally. The existent needs to progress differently, taking what truly matters and composting the rest.

This serves as a consideration for people to take with them—to challenge hierarchies and divisions of labor/specialization. Relatedly, we must realize none of us will be free from the Worldeater and its 'cage system'[4] until everyone is free (see Springer 2019; Bakunin 2005 [1871]). The cage system articulates a prison society, constantly dividing, enclosing, caging and imprisoning all life: trees surrounded by concrete, animals stuck in cages and humans in voluntary and involuntary steel or concrete encasements. There is the need for urgent, brave and unceasing praxis. There is room for theoretical and practical experimentation such as human and non-human urban renewal,[5] which entails disrespecting or, better, re-respecting public and private property in ways that are in accordance with ones' values systems. We want to offer some guidance to more practical and developed bodies of work on this topic. Resonating with Stirner's phantasms and Taussig's devils—and the 'counter-magic' performed by those hostile enough to naming the Worldeater in various vocabularies—the emerging Degrowth movement draws on Castoriadis' (1997) notion of the 'social imaginary' to explain how the 'economy' is discursively born and can be transformed into a lived (en$laving) reality. Similar to the fictions we have discussed as comprising the spirits of power generative of the imperative of total extractivism, Kallis (2018: 23–4) proposes that we 'start reconstituting what an economy is, by asserting its material foundations'. Proceeding to explain the social metabolism of capitalist growth, Kallis (2018) advocating a Degrowth perspective lays out the skeleton(s) for alternative metabolism(s)—'alternative worms' maybe—aimed at auton-

[4] The Worldeater and its technique is built on constantly dividing and caging everything: from enclosing land to proliferating prison architecture as urbanization, not to forget dividing peoples bodies and minds with biomedicine and psychology—even if it has its benefits. Benefits we should keep, but not at the expense of other life forms or the subjugation of other knowledges and methods of healing/living.

[5] This is a play on 'people's urban renewal' from an interview with Mike Davis (2007) to include and support the freedom of non-human life.

omy and conviviality in socio-ecological relationships (see also D'Alisa et al. 2014). This 'counter-magic' that proponents of Degrowth are performing with gusto promises uncovering and reconstituting an oppositional, hostile stance vis-à-vis the capitalist global economy—one that strives for 'life concerned with the beautiful' (Kallis 2018: 35). Proceeding from this, Kallis (2018: 118–23) summarizes nine Degrowth principles: (1) 'an end to exploitation; (2) 'direct democracy'; (3) 'localized production'; (4) 'sharing' and 'reclaiming the commons'; (5) 'shift resources to the provision of relational goods'; (6) 'unproductive expenditures'—concerned with the beautiful; (7) 'care'; (8) 'diverse' economy confining 'production for profit'; and (9) 'decommodification of land, labour and value'. These principles are envisaged as turned into praxis 'from above' through policy interventions and 'from below' through grassroots interventions, the latter proposing means to 'exit' from the (capitalist) economy (Kallis 2018). Moreover, Degrowth implicitly entails radical ecological methods associated with permaculture, forest gardens, 'no-till' gardening (Hemenway 2009; Jacke and Toensmeier 2005; Fukuoka 2010 [1978]) and other non-coercive forms of horticulture long practiced by various Indigenous cultures. This acknowledges that we should be learning from the remaining Indigenous groups still practicing their cultures and subsistence(s) from their habitats. Degrowth, we must add, agreeing with Padini Nirmal and Dianne Rocheleau (2019: 5), must be decolonized, which means submitting itself 'to the actual workings of the living world, rather than trying to control it' and opening itself to a pluriverse of perspectives and knowledges. In these ways, Degrowth promises practical steps out of the entrails and into—back to, forward to—beauty beyond capitalist growth.

This resonates with and complements earlier proposals of bioregionalism (Sale 1991 [1985]), which has slowly spread in popularity (McGinnis 1999; Thayer 2003; Lockyer and Veteto 2013). Bioregionalism is defined as organizing life around 'part of the earth's surface whose rough boundaries are determined by natural characteristics rather than human dictates, distinguishable from other areas by particular attributes of flora, fauna, water, climate, soil, and landforms, and by the human settlements and cultures those attributes have given rise to' (Sale 1991: 162). Flexible to local history, geography and culture, bioregionalism—similar to Degrowth—promotes horizontal political structures, working to nourish habitats, but also avoid predatory economic activity and work-relationships (see Sale 1991; Thayer 2003). The idea is simple, and not without its complications (see Lockyer and Veteto 2013)—such as avoiding pitfalls of xenophobic localism. Yet the issue facing industrial humanity is learning how to live with the land, animals and each other in a respectful and self-supportive way—abolishing the myth of human supremacy—and transforming boredom, loneliness and societal resentment that popularly manifests in suicides, school shootings, murdering animals and joining armed forces. Moreover, bureaucratic institutions, police, imperial war, high-grade (legal) synthetic drugs and psychotropic media bombardment/manipulation through ubiquitous advertisements, television narratives and film reinforce this. The attack against humans and separating them from non-human natures and to have them identify with Worldeater institutions and processes—capitalism—is total and this small book has only charted aspects of this total war waged to regiment vitality into the expanding techniques of the Worldeater.

Degrowth, bioregionalism and other socio-ecological methods serve as practical ways to begin organizing and intervening in our immediate surroundings and institutional policy. We might venture to say that they represent a baseline common vision to be taken up, adapted in the appropriate ways and integrated in variegated broader struggles—to minimize, for example, mass extraction for more electronics, and instead to salvage, recycle and compost what exists into 'appropriate technologies' or appropriate levels of computational technologies. What is advocated here will ideally entail real transition—even though, less optimistically, such transition will more likely result from natural disaster and climate destabilization. Bioregionalism and Decolonial Degrowth deserve greater attention than what is offered here, but we have encouraged exploration and experimentation with these ideas on every level, as the goal is the same. No matter what position people occupy, from being homeless living in the park to politicians (even if one bears greater responsibility): if people want to create a better future, then change your habits, create the socio-political space to allow and support experiments in real socio-ecological harmony. This includes taking direct action and/or joining broader struggles for the survival and creation of a better world than what we know. Corporate activism and the manufacturing of environmental leaders and/or co-optation of movements to create a green economy and (neoliberal) 'climate infrastructure', as revealed by *Wrongkindofgreen.org*, is no substitute. Corporate activism must be subverted and flipped on its head. Profit maximization must cease to exist; an economy based on capital accumulation must be abolished; profiteering from war, housing, health care, water and all the basic necessities of life—must stop. Yes, this may be impossible for many to conceptualize, but the idea is simply to begin—to live life in permanent opposition, to change toward real socio-ecological harmony. The Worldeater stifles this by pushing for institutional strategies of *total integration* and dependence. The real alternative starts with people, their friends—and their children, not the least (see Vetlesen and Willig 2018). Growth must be re-conceptualized; quality of life must be re-evaluated to find what matters, what actually fulfills people and other lifeforms to create mutually supportive environments and relationships. This includes finding out what will stop socio-ecological destruction. Humans reflect their environments and the Worldeater is making sterile, banal, degraded and ultimately suicidal spaces. Call this 'unrealistic', call it utopian, but we live in a once 'impossible' dystopian Science Fiction film and it must be rapidly slowed, stopped, if not reversed to end the disaster of the forming Worldeater.

In line with anarchist insight and Degrowth's practical interventions, we propose that total extractivism demands total liberation in response. Total liberation, argues David Naguib Pellow (2014: 18–9), is an ethos and political stance that has taken prefigurative form among animal rights and radical environmental movements, seen as comprising '(1) an ethic of justice and anti-oppression inclusive of humans, nonhuman animals, and ecosystems; (2) anarchism; (3) anticapitalism; and (4) an embrace of direct action tactics'. While acknowledging the potential for problems and tensions to endure—not picturing this as a flawless ideal—total liberation offers steps toward the necessary integrated approach for human power to assert itself. This has implications for the disciplinary debates covered in this book. Highlighting the hypocrisy of political ecology in its neglect of speciesism or anthroparchy, Springer (2019: 10), explains: 'The infusion of anarchism into political ecology and reading them both alongside more-than human and hybrid geographies offers promise for a more liberatory outlook, whereby "a truly humble, empathic, animal-respecting stance" can begin to emerge'. While total liberation may be regarded as intersectional, it is in fact total. Dubbed the 'ultimate intersectionality', total liberation philosophy transcends specific hierarchies/identity politics (e.g. racism and sexism) and anthropocentrism to offer 'a more all-encompassing idea of intersectionality wherein the goal is confrontation with "The Totality" and total liberation' (Loadenthal 2017: 171, 196). This is the rejection of all domination, linking the liberation of human animals, non-humans animals and the earth to be nothing less than in total conflict with the existent techno-capitalist system. Thus, supportive of Loadenthal (2017: 196), the battle of human animals—queer, white and non-white—are 'inextricably interlinked' as their enemies emanating from a single source of power, which we contend is the Worldeater practice and sprit. Total liberation recognizes the total war of the Worldeater that converts everyone and everything into a commodity—every aspect of life is made legible, commensurable and valuable for confinement and exploitation. Make no mistake, these values are differentiated, hierachialized and changing—the struggle against one power is common, but the situations and oppression faced marks people differently based on gendered, sexual, class and colored lines. While oppression is difficult to quantify, there is greater variation and intensity experienced by different class, genders/non-genders, people (so-called races) and non-humans. Oppression from the techno-capitalist system is not intersectional—it is total. Single issue or reformist visions have a tendency to accept, if not cultivate in whole or in part, the Worldeater system.

Central, however, is how people identify these systematic attacks, relate to governing, market and logistical systems and, most of all, an individual's position in the entrails of the Worldeater. Positionality will make conflicting and contradicting desires, ambitions and real or imagined needs—sometimes making imagined needs real—the longer people incubate in the entrails of capitalist modernity. The point being, now as forever, until all are free, none are free with a cost always being extracted that will come home to roost in the form of political revenge, pests, crop failure, drought and climate change among others. And currently, we are all surrounded by mental and physical cages—the physical reflecting the mental—that are roads gridding out forests and (modernist style) buildings supplanted with great ignorance and carelessness into habitats constructed with reinforced concrete, particle board and complete with door locks and automated key cards. Sometimes we have control over these keys, sometimes it is administrators, other times they are guards—but we live in cages with varying degrees of control that often depends on ones' behavior, civility. There are qualitative—and meaningful—differences, but structurally speaking human and non-human life are saturated with cages, the difference is primarily in the intensity, the layers and saturation of confinement practices. The central difference is that some of the humans call them their own—identifying with their cages—while others do not, and more still suffer from the existence of these infrastructures, even if paradoxically it keeps them warm in the winter. There are, however, other ways to remain warm and live in harmony with our environments and non-human friends that remain undiscovered or neglected by human hubris. We need infrastructure that affirm, restore and expand vital life, not capitalism, not the Worldeater.

The point stands out like few others: Now is the time to return and nourish vitality: from the soil, forests, rivers, air and oceans—to repair reciprocity with the earth and non-human natures—not doubling down at the capitalist casino and systematically devising new technological systems to extract the earth's, and our own, vitality to inject it into industrial and computational systems that are slowly engulfing the earth with the body of the Worldeater. The present trajectory of the techno-capitalist system is precisely on a trajectory for total resource control and, while it is generating record profits, consumer goods and socio-ecological degradation, it appears to be leaving people sick, unfulfilled and dependent. We must abandon the entrails, we must abandon the Worldeater, we must abandon the techno-capitalist system and we must live lives organized around and supporting the vitality that are our ecosystems, our habitats. The benefits of this are underestimated and underaccounted and should remain that way as we begin to live qualitatively and not quantitatively. Remember, the Worldeater is not invincible.

Bibliography

Bakunin, M. (2005 [1871]). God and the State. In D. Guérin (Ed.), *No Masters, No God* (p. 151). Oakland, CA: AK Press.

Castoriadis, C. (1997). *The Imaginary Institution of Society.* Cambridge, MA: MIT Press.

D'Alisa, G., Demaria, F., & Kallis, G. (2014). *Degrowth: A Vocabulary for a New Era.* London: Routledge.

Davis, M. (2007). "Resisting, Subverting and Destroying the Apparatus of Surveillance and Control": An Interview with Mike Davis. *Voices of Resistance from Occupied London, 1*, 16–19.

Dunlap, A. (2014b). Permanent War: Grids, Boomerangs, and Counterinsurgency. *Anarchist Studies, 22*, 55–79.

Escobar, A. (1996). Construction Nature: Elements for a Post-Structuralist Political Ecology. *Futures, 28*, 325–343.

Fukuoka, M. (2010 [1978]). *The One-Straw Revolution: An Introduction to Natural Farming.* New York: New York Review of Books.

Glendinning, C. (1990). *When Technology Wounds: The Human Consequences of Progress.* New York: William Morrow & Co.

Graeber, D. (2018). *Bullshit Jobs: A Theory.* London: Allen Lane.

Hemenway, T. (2009). *Gaia's Garden: A Guide to Home-Scale Permaculture*. White River Junction, VT: Chelsea Green Publishing.
House, B., & Square, T. (2002). High Work Stress Doubles Risk of Cardiovascular Death. *Primary Care, 325*, 6309–6329.
ISIW. (2014). On Deep Green Failures or the Problem of Inert Strategy. *Insurgencies: A Journal on Insurgent Strategy, 1*, 21–56.
Jacke, D., & Toensmeier, E. (2005). *Edible Forest Gardens, Volume II: Ecological Design and Practice for Temperate-Climate Permaculture*. Burlington: Chelsea Green Publishing.
Jensen, D. (2006). *Endgame, Vol. 1: The Problem of Civilisation*. New York: Seven Stories Press.
Kallis, G. (2018). *Degrowth*. New York: Columbia University Press.
Loadenthal, M. (2017). *The Politics of the Attack: Communiqués and Insurrectionary Violence*. Manchester: Manchester University Press.
Lockyer, J., & Veteto, J. R. (2013). *Environmental Anthropology Engaging Ecotopia: Bioregionalism, Permaculture, and Ecovillages*. New York: Berghahn Books.
Matisons, M. R., & Ross, A. R. (2015). *Against Deep Green Resistance*. Retrieved from https://www.counterpunch.org/2015/08/11/against-deep-green-resistance/.
McGinnis, M. V. (1999). *Bioregionalism*. London: Routledge.
Merchant, C. (1983). *The Death of Nature*. New York: Harper & Row.
Nirmal, P., & Rocheleau, D. (2019). Decolonizing Degrowth in the Post-Development Convergence: Questions, Experiences, and Proposals from Two Indigenous Territories. *Environment and Planning E: Nature and Space*, 1–28.
O'Connor, J. (1988). Capitalism, Nature, Socialism: A Theoretical Introduction. *Capitalism Nature Socialism, 1*, 1–38.
Pellow, D. N. (2014). *Total Liberation: The Power and Promise of Animal Rights and the Radical Earth Movement*. Minneapolis, MN and London: University of Minnesota Press.
Romanyshyn, R. (1989). *Technology as Symptom and Dream*. London: Routledge.
Sale, K. (1991 [1985]). *Dwellers in the Land: The Bioregional Vision*. Athens: University of Georgia Press.
Shiva, V. (2002 [1989]). *Staying Alive: Women, Ecology and Development*. London: Zed Books.
Springer, S., Locret, M., Mateer, J., et al. (2019). *Anarchist Political Ecology*. Oakland, CA: PM Press.
Thayer, R. L. (2003). *LifePlace: Bioregional Thought and Practice*. Berkeley, CA: University of California Press.
Vetlesen, A. J., & Willig, R. (2018). *Hva skal vi svare våre barn?* Oslo: Dreyer forlag.

Bibliography

Acosta, A. (2013 [2011]). Extractivism and Neoextractivism: Two Sides of the Same Curse. In M. Lang & D. Mokrani (Eds.), *Beyond Development: Alternative Visions from Latin America* (pp. 61–86). Amsterdam: Transnational Institute.

Adams, R. E. (2014). Natura Urbans, Natura Urbanata: Ecological Urbanism, Circulation, and the Immunization of Nature. *Environment and Planning D: Society and Space, 32*, 12–29.

Agrawal, A. (2005). *Environmentality: Technologies of Government and the Making of Subjects*. Durham, NC: Duke University Press.

Agrawal, A., & Gibson, C. C. (1999). Enchantment and Disenchantment: The Role of Community in Natural Resource Conservation. *World Development, 27*, 629–649.

Agrawal, A., & Redford, K. (2009). Conservation and Displacement: An Overview. *Conservation and Society, 7*, 1–10.

Aguilar-Støen, M. (2016). Beyond Transnational Corporations, Food and Biofuels: The Role of Extractivism and Agribusiness in Land Grabbing in Central America. *Forum for Development Studies, 43*, 155–175.

Akram-Lodhi, A. H. (2018). What Is Critical Agrarian Studies? *Review of African Political Economy*. Retrieved from http://roape.net/2018/03/28/what-is-critical-agrarian-studies/.

Akram-Lodhi, A. H., & Kay, C. (2010a). Surveying the Agrarian Question (Part 1): Unearthing Foundations, Exploring Diversity. *Journal of Peasant Studies, 37*, 177–202.

Akram-Lodhi, A. H., & Kay, C. (2010b). Surveying the Agrarian Question (Part 2): Unearthing Foundations, Exploring Diversity. *Journal of Peasant Studies, 37*, 255–284.

A. Dunlap, J. Jakobsen, *The Violent Technologies of Extraction*,
https://doi.org/10.1007/978-3-030-26852-7

Albertus, M., & Kaplan, O. (2013). Land Reform as a Counterinsurgency Policy: Evidence from Colombia. *Journal of Conflict Resolution, 57*, 198–231.

Alexander, B. K. (2008). *The Globalization of Addiction: A Study in Poverty of the Spirit.* New York: Oxford University Press.

Aman, M., Solangi, K., Hossain, M., et al. (2015). A Review of Safety, Health and Environmental (SHE) Issues of Solar Energy System. *Renewable and Sustainable Energy Reviews, 41*, 1190–1204.

Anderson, K. B. (2010). *Marx at the Margins.* Chicago: University of Chicago Press.

Andreucci, D., García-Lamarca, M., Wedekind, J., et al. (2017). "Value Grabbing": A Political Ecology of Rent. *Capitalism Nature Socialism, 28*, 28–47.

Anthony, P. D. (2001 [1977]). *The Ideology of Work.* Oxon: Tavistock Press/ Routledge.

Arellano-Yanguas, J. (2012). Mining and Conflict in Peru: Sowing the Minerals, Reaping a Hail of Stones. In A. Bebbington (Ed.), *Social Conflict, Economic Development and the Extractive Industry: Evidence from South America* (pp. 89–111). London: Routledge.

Arendt, H. (1962 [1951]). *The Origins of Totalitarianism.* New York: The World Publishing Company.

Arrighi, G. (1994). *The Long Twentieth Century: Money, Power, and the Origins of Our Times.* New York: Verso.

Arsel, M., Hogenboom, B., & Pellegrini, L. (2017). The Extractive Imperative and the Boom in Environmental Conflicts at the End of the Progressive Cycle in Latin America. *The Extractive Industries and Society, 3*(4), 877–879.

Avila, S. (2018). Environmental Justice and the Expanding Geography of Wind Power Conflicts. *Sustainability Science, 13*(3), 599–616.

Bachmann, J., Bell, C., & Holmqvist, C. (2015). *War, Police and Assemblages of Intervention.* Oxon: Routledge.

Baird, I. G., & Barney, K. (2017). The Political Ecology of Cross-Sectoral Cumulative Impacts: Modern Landscapes, Large Hydropower Dams and Industrial Tree Plantations in Laos and Cambodia. *The Journal of Peasant Studies, 44*, 769–795.

Baka, J. (2013). The Political Construction of Wasteland: Governmentality, Land Acquisition and Social Inequality in South India. *Development and Change, 44*(2), 409–428.

Baka, J. (2017). Making Space for Energy: Wasteland Development, Enclosures, and Energy Dispossessions. *Antipode, 49*, 977–996.

Bakunin, M. (2005 [1871]). God and the State. In D. Guérin (Ed.), *No Masters, No God* (p. 151). Oakland, CA: AK Press.

Baldick, C. (1987). *Frankenstein's Shadow: Myth, Monstrosity, and Nineteenth Century Writing.* Oxford: Clarendon Press.

Barbesgaard, M. (2018). Blue Growth: Savior or Ocean Grabbing? *The Journal of Peasant Studies, 45*(1), 130–149.

Bebbington, A. (2012). *Social Conflict, Economic Development and the Extractive Industry: Evidence from South America*. London: Routledge.

Bebbington, A., & Bury, J. (2013). *Subterranean Struggles: New Dynamics of Mining, Oil, and Gas in Latin America*. Austin, TX: University of Texas Press.

Bebbington, A., Bebbington, D. H., Bury, J., et al. (2008). Mining and Social Movements: Struggles Over Livelihood and Rural Territorial Development in the Andes. *World Development, 36*, 2888–2905.

Benjaminsen, T. A., & Bryceson, I. (2012). Conservation, Green/Blue Grabbing and Accumulation by Dispossession in Tanzania. *Journal of Peasant Studies, 39*(2), 335–355.

Berman, E. (1983). *The Ideology of Philanthropy: The Influence of the Carnegie, Ford, and Rockefeller Foundations on American Foreign Policy*. Albany, NY: State University of New York Press.

Bernays, E. (2005 [1928]). *Propaganda, with an Introduction by Michal Crispin*. New York: Ig Publishing.

Bernstein, H. (2010). *Class Dynamics of Agrarian Change*. London: Kumarian Press.

Bernstein, H., Friedmann, H., van der Ploeg, J. D., et al. (2018). Fifty Years of Debate on Peasantries, 1966–2016. *The Journal of Peasant Studies, 45*, 689–714.

BGR (Black and Green Review). (2014–Present). *Black and Green Review*. Retrieved from https://www.blackandgreenreview.org.

Biolsi, T., & Zimmerman, L. J. (1997). *Indians and Anthropologists: Vine Deloria, Jr., and the Critique of Anthropology*. Tucson, AZ: University of Arizona Press.

Birss, M. (2017). Criminalizing Environmental Activism: As Threats to the Environment Increase Across Latin America, New Laws and Police Practices Take Aim Against the Front Line Activists Defending Their Land and Resources. *NACLA Report on the Americas, 49*(3), 315–322.

Black Seed. (2014–Present). Black Seed: A Green Anarchist Publication. Retrieved from http://blackseed.anarchyplanet.org/.

Blaikie, P. (1985). *The Political Economy of Soil Erosion in Developing Countries*. London: Routledge.

Blaikie, P., & Brookfield, H. (1987). *Land Degradation and Society*. London: Methuen.

Blaser, M. (2013). Notes Toward a Political Ontology of 'Environmental' Conflicts. In L. Green (Ed.), *Contested Ecologies: Dialogues in the South on Nature and Knowledge* (pp. 13–27). Cape Town: HSRC Press.

Bonneuil, C., & Fressoz, J.-B. (2016). *The Shock of the Anthropocene: The Earth, History and Us*. New York: Verso Books.

Bookchin, M. (1982). *The Ecology of Freedom: The Emergence and Dissolution of Hierarchy*. Palo Alto, CA: Cheshire Books.

Bookchin, M. (1996). *The Philosophy of Social Ecology: Essays on Dialectical Naturalism*. Montreal, CA: Black Rose Books.

Bookchin, M. (2006). *Social Ecology and Communalism*. Oakland, CA: AK Press.
Boot, M. (2018). *The Road Not Taken: Edward Lansdale and the American Tragedy in Vietnam*. New York: Liveright Publishing.
Borras, S. M., & Franco, J. (2013). Global Land Grabbing and Political Reactions 'From Below'. *Third World Quarterly, 34*, 1723–1747.
Borras, S. M., Jr., Hall, R., Scoones, I., et al. (2011). Towards a Better Understanding of Global Land Grabbing: An Editorial Introduction. *The Journal of Peasant Studies, 38*, 209–216.
Borras, S. M., Jr., Franco, J. C., Isakson, S. R., et al. (2016). The Rise of Flex Crops and Commodities: Implications for Research. *The Journal of Peasant Studies, 43*(1), 93–115.
Borras, S. M., McMichael, P., & Scoones, I. (2010). The Politics of Biofuels, Land and Agrarian Change: Editors' Introduction. *Journal of Peasant Studies, 37*, 575–592.
Borras, S. M., Kay, C., Gómez, S., et al. (2012). Land Grabbing and Global Capitalist Accumulation: Key Features in Latin America. *Canadian Journal of Development Studies/Revue canadienne d'e'tudes du de'veloppement, 33*, 402–416.
Bourdieu, P. (2001). *Masculine Domination*. Cambridge: Polity Press.
Bourgois, P. (2001). The Power of Violence in War and Peace: Post-Cold War Lessons from El Salvador. *Ethnography, 2*, 5–34.
Bridge, G. (2008). Global Production Networks and the Extractive Sector: Governing Resource-Based Development. *Journal of Economic Geography, 8*, 389–419.
Bridge, G. (2014). Resource Geographies II: The Resource-State Nexus. *Progress in Human Geography, 38*(1), 118–130.
Brock, A., & Dunlap, A. (2018). Normalising Corporate Counterinsurgency: Engineering Consent, Managing Resistance and Greening Destruction Around the Hambach Coal Mine and Beyond. *Political Geography, 62*, 33–47.
Brockington, D. (2002). *Fortress Conservation*. Bloomington, IN: Indian University Press.
Brockington, D., & Rosaleen, D. (2010). Antipode Special Issue on Capitalism and Conservation. *Antipode, 42*, 469–484.
Bury, J., & Norris, T. (2013). Rocks, Rangers, and Resistance: Mining and Conservation Frontiers in the Cordillera Huayhuash, Peru. In A. Bebbington & J. Bury (Eds.), *Subterranean Struggles: New Dynamics of Mining, Oil, and Gas in Latin America*. Austin, TX: University of Texas Press.
Büscher, B., & Davidov, V. (2013). *The Ecotourism-Extraction Nexus: Political Economies and Rural Realities of (Un)Comfortable Bedfellows*. London: Routledge.
Büscher, B., & Fletcher, R. (2018). Under Pressure: Conceptualising Political Ecologies of Green Wars. *Conservation and Society, 16*, 105–113.

Büscher, B., & Ramutsindela, M. (2016). Green Violence: Rhino Poaching and the War to Save Southern Africa's Peace Parks. *African Affairs, 115*(458), 1–22.

Büscher, B., & Ramutsindela, M. (2017). Green Violence: Rhino Poaching and the War to Save Southern Africa's Peace Parks. *African Affairs, 115*, 1–22.

Büscher, B., Sullivan, S., Neves, K., et al. (2012). Towards a Synthesized Critique of Neoliberal Conservation. *Capitalism Nature Socialism, 23*, 4–30.

Call, W. (2002). Plan Puebla-Panama. *NACLA Report on the Americas, 35*, 24.

Card, C. (2003). Genocide and Social Death. *Hypatia, 18*, 63–79.

Castellanos-Navarrete, A., & Jansen, K. (2017). Why Do Smallholders Plant Biofuel Crops? The 'Politics of Consent' in Mexico. *Geoforum, 87*(Suppl. C), 15–27.

Castoriadis, C. (1997). *The Imaginary Institution of Society*. Cambridge, MA: MIT Press.

Castro-Gomez, S. (2002). The Social Sciences, Epistemic Violence, and the Problem of the "Invention of the Other". *Nepantla: Views from South, 3*, 269–285.

Cavanagh, C. J., & Benjaminsen, T. A. (2017). Political Ecology, Variegated Green Economies, and the Foreclosure of Alternative Sustainabilities. *Journal of Political Ecology, 24*, 200–216.

Cerdá, I. (1867). *Teoría general de la urbanización, y aplicación de sus principios y doctrinas a la reforma y ensanche de Barcelona*. Madrid: Imprenta Española.

Chamayou, G. (2015 [2013]). *A Theory of the Drone*. New York and London: The New Press.

Chari, S. (2004). *Fraternal Capital: Peasant-Workers, Self-Made Men, and Globalization in Provincial India*. Stanford, CA: Stanford University Press.

Chigumira, E. (2018). Political Ecology of Agrarian Transformation: The Nexus of Mining and Agriculture in Sanyati District, Zimbabwe. *Journal of Rural Studies, 61*, 265–276.

Churchill, W., & Orelus, P. (2012). Confronting Western Colonialism, American Racism, and White Supremacy: Ward Churchill and Pierre Orelus in Dialogue. In P. W. Orelus (Ed.), *A Decolonizing Encounter: Ward Churchill and Antonia Darder in Dialogue* (pp. 56–112). New York: Peter Lang.

Clark, J., & Martin, C. (Eds.). (2013). *Anarchy, Geography, Modernity: Selected Writings of Elisée Reclus*. Oakland, CA: PM Press.

Clastres, P. (1989 [1974]). *Society Against the State: Essays in Political Anthropology*. New York: Zone Books.

Clastres, P., Hurley, R., & Stein, A. (1977). *Society Against the State*. Oxford: Blackwell.

Collard, R.-C. (2015). Ethics in Research Beyond the Human. In T. Perreault et al. (Eds.), *The Routledge Handbook of Political Ecology* (pp. 127–139). New York: Routledge.

Copeland, N. (2012). Greening the Counterinsurgency: The Deceptive Effects of Guatemala's Rural Development Plan of 1970. *Development and Change, 43*, 975–998.

Correa-Cabrera, G. (2017). *Los Zetas Inc.: Criminal Corporations, Energy, and Civil War in Mexico*. Austin, TX: University of Texas Press.

Corson, C., MacDonald, K. I., & Neimark, B. (2013). Grabbing "Green": Markets, Environmental Governance and the Materialization of Natural Capital. *Human Geography, 6*, 1–15.

Cowan, T. (2018). The Urban Village, Agrarian Transformation, and Rentier Capitalism in Gurgaon, India. *Antipode, 50*, 1244–1266.

Cowen, D. (2014). *The Deadly Life of Logistics*. Minneapolis, MN: University of Minnesota Press.

CPI (Communist Party of India) Maoist. (2004). *Party Programme. Central Committee (P)*. Retrieved from http://www.bannedthought.net/India/CPI-Maoist-Docs/.

Cramer, C., & Richards, P. (2011). Violence and War in Agrarian Perspective. *Journal of Agrarian Change, 11*, 277–297.

Crosby, A. C., & Monaghan, J. (2018). *Policing Indigenous Movements: Dissent and the Security State*. Black Point: Fernwood Publishing.

Cullather, N. (2006). "The Target Is the People": Representations of the Village in Modernization and U.S. National Security Doctrine. *Cultural Politics, 2*, 29–48.

Cullather, N. (2013 [2010]). *The Hungry World: America's Cold War Battle Against Poverty in Asia*. Cambridge: Harvard University Press.

D'Alisa, G., Demaria, F., & Kallis, G. (2014). *Degrowth: A Vocabulary for a New Era*. London: Routledge.

Dalby, S. (2015). Geoengineering: The Next Era of Geopolitics? *Geography Compass, 9*, 190–201.

Das, V., & Poole, D. (2004). *Anthropology in the Margins of the State*. Santa Fe: SAR Press.

Davis, M. (2007). "Resisting, Subverting and Destroying the Apparatus of Surveillance and Control": An Interview with Mike Davis. *Voices of Resistance from Occupied London, 1*, 16–19.

Davis, R., & Zannis, M. (1973). *The Genocide Machine in Canada*. Montreal, CA: Black Rose Books.

Demaria, F., & Kothari, A. (2017). The Post-Development Dictionary Agenda: Paths to the Pluriverse. *Third World Quarterly, 38*, 2588–2599.

Derrida, J. (2012 [1993]). *Specters of Marx: The State of the Debt, the Work of Mourning and the New International*. London: Routledge.

DESA U. (2018a). *2018 Revision of World Urbanization Prospects*. Retrieved from https://www.un.org/development/desa/publications/2018-revision-of-world-urbanization-prospects.html.

DESA U. (2018b). *2018 Energy Statistics Pocketbook*. Retrieved from https://unstats.un.org/unsd/energy/pocket/2018/2018pb-web.pdf.

Devine, J. (2014). Counterinsurgency Ecotourism in Guatemala's Maya Biosphere Reserve. *Environment and Planning D: Society and Space, 32*, 984–1001.

Dıez, M., Alvarez, R., & Barriocanal, C. (2002). Coal for Metallurgical Coke Production: Predictions of Coke Quality and Future Requirements for Cokemaking. *International Journal of Coal Geology, 50*, 389–412.

Downey, L., Bonds, E., & Clark, K. (2010). Natural Resource Extraction, Armed Violence, and Environmental Degradation. *Organization Environment, 23*, 453–474.

Duffield, M. (2011). Total War as Environmental Terror: Linking Liberalism, Resilience, and the Bunker. *The South Atlantic Quarterly, 110*, 757–769.

Duffy, R. (2015). Nature-Based Tourism and Neoliberalism: Concealing Contradictions. *Tourism Geographies, 17*, 529–543.

Dugger, W. (1989). *Corporate Hegemony*. Westport, CT: Greenwood Press.

Dunlap, A. (2014a). Power: Foucault, Dugger and Social Warfare. In B. Collective (Ed.), *The BASTARD Chronicles: Social War* (pp. 55–106). Berkeley, CA: Ardent Press.

Dunlap, A. (2014b). Permanent War: Grids, Boomerangs, and Counterinsurgency. *Anarchist Studies, 22*, 55–79.

Dunlap, A. (2016). Counter-Insurgency: Let's Remember Where Prevention Comes from and Its Implications. *Critical Studies on Terrorism, 9*, 380–384.

Dunlap, A. (2017a). The Town Is Surrounded': From Climate Concerns to Life Under Wind Turbines in La Ventosa, Mexico. *Human Geography, 10*, 16–36.

Dunlap, A. (2017b). Wind Energy: Toward a "Sustainable Violence" in Oaxaca, Mexico. *NACLA, 49*, 483–488.

Dunlap, A. (2018a). Counterinsurgency for Wind Energy: The Bíi Hioxo Wind Park in Juchitán, Mexico. *The Journal of Peasant Studies, 45*, 630–652.

Dunlap, A. (2018b). Insurrection for Land, Sea and Dignity: Resistance and Autonomy Against Wind Energy in Álvaro Obregón, Mexico. *Journal of Political Ecology, 25*, 120–143.

Dunlap, A. (2018c). The 'Solution' Is Now the 'Problem': Wind Energy, Colonization and the 'Genocide-Ecocide Nexus' in the Isthmus of Tehuantepec, Oaxaca. *The International Journal of Human Rights, 42*, 550–573.

Dunlap, A. (2018d). "A Bureaucratic Trap": Free, Prior and Informed Consent (FPIC) and Wind Energy Development in Juchitán, Mexico. *Capitalism Nature Socialism, 29*, 88–108.

Dunlap, A. (2018e). *Reconsidering the Logistics of Autonomy: Ecological Autonomy, Self-Defense and the Polícia Comunitaria in Álvaro Obregón*, Mexico. Retrieved from https://www.tni.org/files/article-downloads/erpi_cp_8_dunlap.pdf.

Dunlap, A. (2018f). *End the "Green" Delusions: Industrial-Scale Renewable Energy Is Fossil Fuel+*. Retrieved from https://www.versobooks.com/blogs/3797-end-the-green-delusions-industrial-scale-renewable-energy-is-fossil-fuel.

Dunlap, A. (2018g). Book Review: The Anarchist Roots of Geography: Toward Spatial Emancipation by Simon Springer. *Human Geography, 11*, 62–64.

Dunlap, A. (2018h). *Resisting Development: The Politics of the Zad and NoTav*. Retrieved from https://entitleblog.org/2018/09/06/the-politics-of-the-zad-and-notav/.

Dunlap, A. (2019a). Revisiting the Wind Energy Conflict in Gui'Xhi' Ro/Álvaro Obregón: Interview with an Indigenous Anarchist. *Journal of Political Ecology, 26*, 150–166.

Dunlap, A. (2019b). 'Agro sí, mina NO!' The Tía Maria Copper Mine, State Terrorism and Social War by Every Means in the Tambo Valley, Peru. *Political Geography, 71*, 10–25.

Dunlap, A. (2019c). *Renewing Destruction: Wind Energy Development in Oaxaca, Mexico*. London: Rowman & Littlefield.

Dunlap, A. (forthcoming). Wind, Coal and Copper: The Politics of Land Grabbing, Counterinsurgency and the Social Engineering of Extraction. Globalizations.

Dunlap, A., & Brock, A. (2019). When the Wolf Guards the Sheep: Green Extractivism in Germany and Mexico. In S. Springer, M. Locret, J. Mateer, et al. (Eds.), *Anarchist Political Ecology* (Vol. 3). Oakland, CA: PM Press.

Dunlap, A., & Fairhead, J. (2014). The Militarisation and Marketisation of Nature: An Alternative Lens to 'Climate-Conflict'. *Geopolitics, 19*, 937–961.

Dunlap, A., & Sullivan, S. (2019). A Faultline in Neoliberal Environmental Governance Scholarship? Or, Why Accumulation-by-Alienation Matters. *Environment & Planning E: Nature and Space*, 1–19.

Edelman, M., & Wolford, W. (2017). Introduction: Critical Agrarian Studies in Theory and Practice. *Antipode, 49*, 959–976.

Edelman, M., Oya, C., & SMB, J. (2013). Global Land Grabs: Historical Processes, Theoretical and Methodological Implications and Current Trajectories. *Third World Quarterly, 34*, 1517–1531.

Ehrnström-Fuentes, M., & Kröger, M. (2018). Birthing Extractivism: The Role of the State in Forestry Politics and Development in Uruguay. *Journal of Rural Studies, 57*, 197–208.

Elden, S. (2010). Land, Terrain, Territory. *Progress in Human Geography, 34*(6), 799–817.

Eldridge, E. R., & Reinke, A. J. (2018). Introduction: Ethnographic Engagement with Bureaucratic Violence. *Conflict and Society, 4*, 94–98.

Ellul, J. (1964 [1954]). *The Technological Society*. New York: Vintage Books.

Ellul, J. (1980). *The Technological System*. New York: Continuum.

Emel, J., & Neo, H. (Eds.). (2015). *Political Ecologies of Meat*. London: Routledge.

Enns, C. (2019). Infrastructure Projects and Rural Politics in Northern Kenya: The Use of Divergent Expertise to Negotiate the Terms of Land Deals for Transport Infrastructure. *The Journal of Peasant Studies, 46*, 358–376.

Enzensberger, H. M. (1974). A Critique of Political Ecology. *New Left Review, 84*, 3–32.
Escobar, A. (1996). Construction Nature: Elements for a Post-Structuralist Political Ecology. *Futures, 28*, 325–343.
Escobar, A. (2006). Difference and Conflict in the Struggle over Natural Resources: A Political Ecology Framework. *Development, 49*, 6–13.
Escobar, A. (2008). *Territories of Difference: Place, Movements, Life, Reads.* Durham: Duke University Press.
Escobar, A. (2012 [1995]). *Encountering Development: The Making and Unmaking of the Third World.* Princeton, NJ: Princeton University Press.
Escobar, A. (2018). *Designs for the Pluriverse: Radical Interdependence, Autonomy, and the Making of Worlds.* Durham, NC: Duke University Press.
Esteva, G. (2014). Commoning in the New Society. *Community Development Journal, 49*, 144–159.
Esteva, G., & Prakash, M. S. (2014). *Grassroots Postmodernism: Remaking the Soil of Cultures.* London: Zed Books.
EZLN. (2016). *Critical Thought in the Face of the Capitalist Hydra: I: Contributions by the Sixth Commission of the EZLN.* Brisbane: Paper Boat Press.
Fairbairn, M., Fox, J., Isakson, S. R., et al. (2014). Introduction: New Directions in Agrarian Political Economy. *Journal of Peasant Studies, 41*, 653–666.
Fairhead, J., Leach, M., & Scoones, I. (2012). Green Grabbing: A New Appropriation of Nature? *Journal of Peasant Studies, 39*, 237–261.
Federici, S. (2009 [2004]). *Caliban and the Witch: Women, the Body and Primitive Accumulation.* New York: Autonomedia.
Finley-Brook, M., & Thomas, C. (2011). Renewable Energy and Human Rights Violations: Illustrative Cases from Indigenous Territories in Panama. *Annals of the Association of American Geographers, 101*, 863–872.
Fitzpatrick, B. (2018). *Corrosive Consciousness.* Jacksonville, FL: Enemy Combatant.
Fletcher, R. (2010). Neoliberal Environmentality: Towards a Poststructuralist Political Ecology of the Conservation Debate. *Conservation and Society, 8*, 171–181.
Fletcher, R. (2017). Environmentality Unbound: Multiple Governmentalities in Environmental Politics. *Geoforum, 85*, 311–315.
Fletcher, R., Dressler, W. H., Anderson, Z. R., et al. (2018). Natural Capital Must Be Defended: Green Growth as Neoliberal Biopolitics. *The Journal of Peasant Studies, 46*, 1–28.
FM3-24. (2014). Insurgencies and Countering Insurgencies. Retrieved from http://fas.org/irp/doddir/army/fm3-24.pdf.
Foucault, M. (1995 [1977]). *Discipline and Punish: The Birth of the Prison.* New York: Random House.

Foucault, M. (1998 [1978]). *The Will to Knowledge: The History of Sexuality: 1.* London: Penguin Books.

Foucault, M. (2003 [1997]). *"Society Must Be Defended": Lectures at the College de France 1975–1976.* New York: Picador.

Foucault, M. (2007a [1978]). *Security, Territory, Population: Lectures at the College de France 1977–1978.* New York: Picador.

Foucault, M. (2007b [1976]). The Meshes of Power. In J. W. Crampton & S. Elden (Eds.), *Space, Knowledge and Power: Foucault and Geography.* Farnham: Ashgate Publishing Limited.

Franco, J. (2014). *Reclaiming Free Prior and Informed Consent (FPIC) in the Context of Global Land Grabs.* Transnational Institute for Hand off the Land Alliance.

Franco, J. C., & Borras, S. M. (2019). Grey Areas in Green Grabbing: Subtle and Indirect Interconnections Between Climate Change Politics and Land Grabs and Their Implications for Research. *Land Use Policy, 84*, 192–199.

Franquesa, J. (2018). *Power Struggles: Dignity, Value, and the Renewable Energy Frontier in Spain.* Bloomington, IN: Indiana University Press.

Fukuoka, M. (2010 [1978]). *The One-Straw Revolution: An Introduction to Natural Farming.* New York: New York Review of Books.

GA. (2005). What Is Green Anarchy? An Introduction to Anti-Civilization Anarchist Thought and Practice. Retrieved from https://theanarchistlibrary.org/library/anonymous-what-is-green-anarchy.

GA. (2012). *Uncivilized: The Best of Green Anarchy.* Berkeley, CA: Ardent Press.

GA (Green Anarchy). (2002–2009). Green Anarchy: An Anti-Civilization. *Journal of Theory and Action.* Retrieved from http://greenanarchy.anarchyplanet.org/.

Gago, V., & Mezzadra, S. (2017). A Critique of the Extractive Operations of Capital: Toward an Expanded Concept of Extractivism. *Rethinking Marxism, 29*, 574–591.

Galeano, E. (1997 [1973]). *Open Veins of Latin America: Five Centuries of the Pillage of a Continent.* London: Monthly Review Press.

Galloway, A. R. (2014). The Cybernetic Hypothesis. *Differences, 25*, 107–131.

Galtung, J. (1969). Violence, Peace, and Peace Research. *Journal of Peace Research, 6*, 167–191.

Gamu, J. K., & Dauvergne, P. (2018). The Slow Violence of Corporate Social Responsibility: The Case of Mining in Peru. *Third World Quarterly, 39*, 959–975.

Geenen, S., & Verweijen, J. (2017). Explaining Fragmented and Fluid Mobilization in Gold Mining Concessions in Eastern Democratic Republic of the Congo. *The Extractive Industries and Society, 4*(4), 758–765.

Gelderloos, P. (2013). *The Failure of Nonviolence: From Arab Spring to Occupy.* Seattle, WA: Left Bank Books.

Gelderloos, P. (2017). *Worshiping Power: An Anarchist View of Early State Formation*. Oakland, CA: AK Press.
Gerber, J.-F. (2011). Conflicts Over Industrial Tree Plantations in the South: Who, How and Why? *Global Environmental Change, 21*, 165–176.
Giarracca, N., & Teubal, M. (2014). Argentina: Extractivist Dynamics of Soy Production and Open-Pit Mining. In H. Veltmeyer & J. F. Petras (Eds.), *The New Extractivism: A Post-Neoliberal Development Model or Imperialism of the Twenty-First Century?* (pp. 47–79). London: Zed Books.
Glendinning, C. (1990). *When Technology Wounds: The Human Consequences of Progress*. New York: William Morrow & Co.
Gompert, D. C., Grissom, J. G., IV, et al. (2008). *War by Other Means: Building Complete and Balanced Capabilities for Counterinsurgency*. RAND Counterinsurgency Study Final Report. Santa Monica RAND Corporation National Security Research Division.
González-Hidalgo, M., & Zografos, C. (2017). How Sovereignty Claims and "Negative" Emotions Influence the Process of Subject-Making: Evidence from a Case of Conflict Over Tree Plantations from Southern Chile. *Geoforum, 78*, 61–73.
Gorrion, A. (2013). Anarchy in World Systems. *The Anvil Review*. Retrieved from http://theanvilreview.org/print/anarchy-in-world-systems/.
Graeber, D. (2012). *Debt: The First 5,000 Years*. London: Melville House.
Graeber, D. (2018). *Bullshit Jobs: A Theory*. London: Allen Lane.
Graham, S. (2011). *Cities Under Siege: The New Military Urbanism*. New York: Verso.
Grajales, J. (2013). State Involvement, Land Grabbing and Counter-Insurgency. *Development and Change, 44*, 211–232.
Gudynas, E. (2009). Diez tesis urgentes sobre el nuevo extrac-tivismo. In J. Schuldt, A. Acosta, A. Barandiarán, et al. (Eds.), *Extractivismo, política y sociedad*. Quito: Centro An-dino de Acción Popular (CAAP) and Centro Latinoamericano de Ecología Social (CLAES).
Gudynas, E. (2010). The New Extractivism of the 21st Century: Ten Urgent Theses About Extractivism in Relation to Current South American Progressivism. *Americas Program Report, 21*, 1–14.
Gudynas, E. (2013 [2011]). Debates on Development and Its Alternatives in Latin America: A Brief Heterodox Guide. In M. Lang & D. Mokrani (Eds.), *Beyond Development: Alternative Visions from Latin America* (pp. 15–40). Amsterdam: Transnational Institute.
Gudynas, E. (2017). Post-Development and Other Critiques of the Roots of Development. In *The Essential Guide to Critical Development Studies* (pp. 104–113). London: Routledge.
Guezuraga, B., Zauner, R., & Pölz, W. (2012). Life Cycle Assessment of Two Different 2 MW Class Wind Turbines. *Renewable Energy, 37*, 37–44.

Gunderson, R. (2011). From Cattle to Capital: Exchange Value, Animal Commodification, and Barbarism. *Critical Sociology, 39*, 259–275.

Hadnagy, C. (2011). *Social Engineering: The Art of Human Hacking*. Indianapolis, IN: Wiley Publishing.

Halberstam, D. (1972 [1969]). *The Best and the Brightest*. New York: Random House.

Hall, D. (2011). Land Grabs, Land Control, and Southeast Asian Crop Booms. *The Journal of Peasant Studies, 38*, 837–857.

Hall, D., Hirsch, P., & Li, T. M. (2011). *Powers of Exclusion: Land Dilemmas in Southeast Asia*. Honolulu: University of Hawaii Press.

Hall, R., Edelman, M., Borras, S. M., Jr., Scoones, I., White, B., & Wolford, W. (2015). Resistance, Acquiescence or Incorporation? An Introduction to Land Grabbing and Political Reactions 'From Below'. *Journal of Peasant Studies, 42*(3–4), 467–488.

Han, C. (2017). *Psycho-Politics*. London: Verso.

Haraway, D. (2015). Anthropocene, Capitalocene, Plantationocene, Chthulucene: Making Kin. *Environmental Humanities, 6*(1), 159–165.

Hardin, G. (1968). The Tragedy of the Commons. *Science, 162*, 1234–1248.

Harrison, F. V. (1991). *Decolonizing Anthropology Moving Further Toward an Anthropology for Liberation*. Washington, DC: American Anthropological Association.

Harvey, D. (1974). Ideology and Population Theory. *International Journal of Health Services, 4*(3), 515–537.

Harvey, D. (2003). *The New Imperialism*. New York: Oxford University Press.

Harvey, D. (2018 [1982]). *The Limits to Capital*. London: Verso.

Hedegaard, H., Curtin, S. C., & Warner, M. (2018). *Suicide Rates in the United States Continue to Increase*. Retrieved from https://www.cdc.gov/nchs/products/databriefs/db309.htm.

Hemenway, T. (2009). *Gaia's Garden: A Guide to Home-Scale Permaculture*. White River Junction, VT: Chelsea Green Publishing.

Heynen, N., & Robbins, P. (2005). The Neoliberalization of Nature: Governance, Privatization, Enclosure and Valuation. *Capitalism Nature Socialism, 16*, 5–8.

Heynen, N., & Sant, L. V. (2015). Political Ecologies of Activism and Direct Action Politics. In T. Perreault, G. Bridge, & J. McCarthy (Eds.), *The Routledge Handbook of Political Ecology*. London: Routledge.

Heynen, N., Kaika, M., & Swyngedouw, E. (Eds.). (2006). *The Nature of Cities: Urban Political Ecology and the Politics of Urban Metabolism*. New York: Routledge.

Hickel, J. L., & Kallis, G. (2019). Is Green Growth Possible? *New Political Economy*. https://doi.org/10.1080/13563467.2019.1598964.

Hildyard, N. (2016). *Licensed Larceny: Infrastructure, Financial Extraction and the Global South*. Manchester: Manchester University Press.

Hill, W., Byrne, J., & de Vasconcellos Pegas, F. (2016). The Ecotourism-Extraction Nexus and Its Implications for the Long-Term Sustainability of Protected Areas: What Is Being Sustained and Who Decides? *Journal of Political Ecology, 23*, 308–327.

Hindery, D. (2013). *From Enron to Evo: Pipeline Politics, Global Environmentalism, and Indigenous Rights in Bolivia*. Tucson, AZ: University of Arizona Press.

Hinton, A. L. (2002). *Annihilating Difference: The Anthropology of Genocide*. Berkeley: University of California Press.

Hoenderdaal, S., Espinoza, L. T., Marscheider-Weidemann, F., et al. (2013). Can a Dysprosium Shortage Threaten Green Energy Technologies? *Energy, 49*, 344–355.

Holmes, G. (2014). What Is a Land Grab? Exploring Green Grabs, Conservation, and Private Protected Areas in Southern Chile. *Journal of Peasant Studies, 41*, 547–567.

House, B., & Square, T. (2002). High Work Stress Doubles Risk of Cardiovascular Death. *Primary Care, 325*, 6309–6329.

Hribal, J. (2010). *Fear of the Animal Planet: The Hidden History of Animal Resistance*. Petrolia and Oakland, CA: CounterPunch and AK Press.

Huber, M. T. (2009). Energizing Historical Materialism: Fossil Fuels, Space and the Capitalist Mode of Production. *Geoforum, 40*, 105–115.

Huff, A., & Brock, A. (2017). *Intervention—"Accumulation by Restoration: Degradation Neutrality and the Faustian Bargain of Conservation Finance"*. Retrieved from https://antipodefoundation.org/2017/11/06/accumulation-by-restoration/.

Hunsberger, C., Corbera, E., Borras, S. M., Jr., et al. (2017). Climate Change Mitigation, Land Grabbing and Conflict: Towards a Landscape-Based and Collaborative Action Research Agenda. *Canadian Journal of Development Studies/Revue canadienne d'études du développement, 38*, 305–324.

Igoe, J. (2010). The Spectacle of Nature in the Global Economy of Appearances: Anthropological Engagements with the Spectacular Mediations of Transnational Conservation. *Critique of Anthropology, 30*, 375–397.

Illich, I. (1970 [1969]). *Planned Poverty: The End Result of Technical Assistance. In Celebration of Awareness: A Call for Institutional Revolution*. London: Marion Boyars.

Illich, I. (1972). *Limits to Medicine: Medical Nemesis—The Expropriation of Health*. London: Marion Boyars.

Illich, I. (1973a). *Energy and Equity*. London: Calder & Boyars Ltd.

Illich, I. (1973b). *Tools for Conviviality*. London: Calder & Boyars.

Illich, I. (1978a). *The Right to Useful Unemployment and Its Professional Enemies*. London: Marion Boyars.

Illich, I. (1978b). *Towards a History of Needs*. New York: Pantheon Books.

Illich, I. (2002 [1970]). *Deschooling Society*. London: Calder & Boyars Ltd.

IPBES. (2019). Report of the Plenary of the Intergovernmental Science-Policy Platform on Biodiversity and Ecosystem Services on the Work of Its Seventh Session [Advance Version]. Retrieved from https://www.ipbes.net/system/tdf/ipbes_7_10_add-1-_advance.pdf?file=1&type=node&id=35329.

ISIW (The Institute for the Study of Insurgent Warfare). (2014). On Deep Green Failures or the Problem of Inert Strategy. *Insurgencies: A Journal on Insurgent Strategy, 1*, 21–56.

Jacke, D., & Toensmeier, E. (2005). *Edible Forest Gardens, Volume II: Ecological Design and Practice for Temperate-Climate Permaculture*. Burlington: Chelsea Green Publishing.

Jakhu, R. S., et al. (2017). *Space Mining and Its Regulation*. Chichester: Springer.

Jakobsen, J. (2016). Disappearing Landlords and the Unmaking of Revolution: Maoist Mobilization, the State and Agrarian Change in Northern Telangana. In K. B. Nielsen & A. G. Nilsen (Eds.), *Social Movements and the State in India: Deepening Democracy?* (pp. 239–267). London: Palgrave Macmillan.

Jakobsen, J. (2017). Inside and Outside the State: Imagining the Indian State Among Subalternists and Naxalites. In P. Basu (Ed.), *Naxalite Politics: Post-Structuralist, Postcolonial and Subaltern Perspectives* (pp. 283–299). Kolkata: Setu Prakashani.

Jakobsen, J. (2018a). Neoliberalising the Food Regime 'Amongst Its Others': The Right to Food and the State in India. *The Journal of Peasant Studies*, 1–21. https://doi.org/10.1080/03066150.2018.1449745.

Jakobsen, J. (2018b). Towards a Gramscian Food Regime Analysis of India's Agrarian Crisis: Counter-Movements, Petrofarming and Cheap Nature. *Geoforum, 90*, 1–10.

Jakobsen, J., & Hansen, A. (2019). Geographies of Meatification: An Emerging Asian Meat Complex. *Globalizations*, 1–17. https://doi.org/10.1080/14747731.2019.1614723.

Jakobsen, J., & Nielsen, K. B. (forthcoming). Compounding Aspirations: Grounding Hegemonic Processes in India's Rural Transformations. *Canadian Journal of Development Studies/Revue canadienne d'études du développement.*

Jensen, D. (2006). *Endgame, Vol. 1: The Problem of Civilisation*. New York: Seven Stories Press.

Jensen, D. (2016). *The Myth of Human Supremacy*. New York and Oakland, CA: Seven Stories Press.

Kallis, G. (2018). *Degrowth*. New York: Columbia University Press.

Kautsky, K. (1988 [1899]). *The Agrarian Question* (Vol. 2). London: Zwan Publications.

Kelly, A. (2011). Conservation Practice as Primitive Accumulation. *The Journal of Peasant Studies, 38*, 683–701.

Khalili, L. (2015). Counterterrorism and Counterinsurgency in the Neoliberal Age. In A. Ghazal & J. Hanssen (Eds.), *Oxford Handbook of Contemporary Middle-Eastern and North African History*. Oxford: Oxford University Press.

Kiezebrink, V., Wilde-Ramsing, J., & Kate Gt. (2018). *Human Rights in Wind Turbine Supply Chains: Towards a Truly Sustainable Energy Transition*. Retrieved from https://www.somo.nl/wp-content/uploads/2018/01/Final-ActionAid_Report-Human-Rights-in-Wind-Turbine-Supply-Chains.pdf.

Kirsch, S. (2014). *Mining Capitalism: The Relationship Between Corporations and Their Critics*. Berkeley, CA: University of California Press.

Kleinman, A. (2000). The Violences of Everyday Life. In V. Das, A. Kleinman, M. Ramphele, et al. (Eds.), *Violence and Subjectivity* (pp. 226–241). Berkeley, CA: University of California Press.

Kröger, M. (2016a). Spatial Causalities in Resource Rushes: Notes from the Finnish Mining Boom. *Journal of Agrarian Change, 16*, 543–570.

Kröger, M. (2016b). The Political Economy of 'Flex Trees': A Preliminary Analysis. *The Journal of Peasant Studies, 43*(4), 886–909.

Lahiri-Dutt, K. (2018a). Extractive Peasants: Reframing Informal Artisanal and Small-Scale Mining Debates. *Third World Quarterly, 39*, 1561–1582.

Lahiri-Dutt, K. (Ed.). (2018b). *Between the Plough and the Pick: Informal Mining in the Contemporary World*. Canberra: Australian National University Press.

Landstreicher, W. (2009). *Willful Disobedience*. Berkeley, CA: Ardent Press.

Lang, M., & Mokrani, D. (2013 [2011]). *Beyond Development: Alternative Visions from Latin America*. Amsterdam: Transnational Institute.

Le Billon, P. (2001). The Political Ecology of War: Natural Resources and Armed Conflicts. *Political Geography, 20*, 561–584.

Le Billon, P. (2012). *Wars of Plunder: Conflicts, Profits and the Politics of Resources*. New York: Columbia University Press.

Lenin, V. I. (1964 [1899]). *The Development of Capitalism in Russia*. Moscow: Progress Publishers.

Lerche, J., Shah, A., & Harriss-White, B. (2013). Introduction: Agrarian Questions and Left Politics in India. *Journal of Agrarian Change, 13*, 337–350.

Levien, M. (2018). *Dispossession Without Development: Land Grabs in Neoliberal India*. New York: Oxford University Press.

Levien, M., Watts, M., & Yan, H. (2018). Agrarian Marxism. *The Journal of Peasant Studies, 45*(5–6), 853–888.

Li, T. M. (2009). Exit from Agriculture: A Step Forward or a Step Backward for the Rural Poor? *The Journal of Peasant Studies, 36*, 629–636.

Li, T. M. (2011). Centering Labor in the Land Grab Debate. *Journal of Peasant Studies, 38*, 281–298.

Li, T. M. (2014). *Land's End: Capitalist Relations on an Indigenous Frontier*. Durham, NC: Duke University Press.

Li, T. M. (2018). After the Land Grab: Infrastructural Violence and the "Mafia System" in Indonesia's Oil Palm Plantation Zones. *Geoforum, 96*, 328–337.

Light, J. S. (2003). *From Warfare to Welfare: Defense Intellectuals and Urban Problems in Cold War America*. London: The John Hopkins University Press.

Loadenthal, M. (2017). *The Politics of the Attack: Communiqués and Insurrectionary Violence*. Manchester: Manchester University Press.

Lockyer, J., & Veteto, J. R. (2013). *Environmental Anthropology Engaging Ecotopia: Bioregionalism, Permaculture, and Ecovillages*. New York: Berghahn Books.

Loftus, A. (2013). Gramsci, Nature and the Philosophy of Praxis. In M. Ekers, G. Hart, S. Kipfer, & A. Loftus (Eds.), *Gramsci: Space, Nature, Politics* (pp. 178–197). Oxford: Wiley-Blackwell.

Loftus, A. (2018). Political Ecology II: Whither the State? *Progress in Human Geography*, 1–11.

Lund, C. (2016). Rule and Rupture: State Formation Through the Production of Property and Citizenship. *Development and Change, 47*, 1199–1228.

Lund, C. (2018). Predatory Peace. Dispossession at Aceh's Oil Palm Frontier. *The Journal of Peasant Studies, 45*, 431–452.

Lunstrum, E. (2014). Green Militarization: Anti-Poaching Efforts and the Spatial Contours of Kruger National Park. *Annals of the Association of American Geographers, 104*, 816–832.

Machlis, G. E., & Hanson, T. (2008). Warfare Ecology. *BioScience, 58*(8), 729–736.

Maldonado-Torres, N. (2008). *Against War*. Durham, NC: Duke University Press.

Malm, A. (2016). *Fossil Capital: The Rise of Steam Power and the Roots of Global Warming*. London: Verso Books.

Mann, G. (2009). Should Political Ecology Be Marxist? A Case for Gramsci's Historical Materialism. *Geoforum, 40*(3), 335–344.

Mann, G., & Wainwright, J. (2018). *Climate Leviathan*. London: Verso Books.

Marijnen, E. (2017). The 'Green Militarisation' of Development Aid: The European Commission and the Virunga National Park, DR Congo. *Third World Quarterly, 38*, 1566–1582.

Marijnen, E., & Verweijen, J. (2016). Selling Green Militarization: The Discursive (Re)Production of Militarized Conservation in the Virunga National Park, Democratic Republic of the Congo. *Geoforum, 75*, 274–285.

Marker, M. (2003). Indigenous Voice, Community, and Epistemic Violence: The Ethnographer's "Interests" and What "Interests" the Ethnographer. *International Journal of Qualitative Studies in Education, 16*, 361–375.

Marston, A., & Perreault, T. (2017). Consent, Coercion and Cooperativismo: Mining Cooperatives and Resource Regimes in Bolivia. *Environment and Planning A: Economy and Space, 49*(2), 252–272.

Martínez-Alier, J. (2002). *The Environmentalism of the Poor: A Study of Ecological Conflicts and Valuation*. Northampton: Edward Elgar.

Martínez-Alier, J., & O'Connor, M. (1995). Ecological and Economic Distribution Conflicts. In R. Costanza, O. Segura, & J. Martínez-Alier (Eds.), *Getting Down to Earth: Practical Applications of Ecological Economics* (pp. 153–183). Washington, DC: Island Press.

Martinez-Alier, J., & Walter, M. (2016). Social Metabolism and Conflicts Over Extractivism. In F. De Castro, B. Hogenboom, & M. Baud (Eds.), *Environmental Governance in Latin America*. Hampshire: Palgrave Macmillan.

Martinez-Alier, J., Temper, L., Del Bene, D., et al. (2016). Is There a Global Environmental Justice Movement? *The Journal of Peasant Studies, 43*, 731–755.

Martínez-Torres, M. E., & Rosset, P. M. (2010). La Vía Campesina: The Birth and Evolution of a Transnational Social Movement. *The Journal of Peasant Studies, 37*, 149–175.

Marx, K. (1982 [1867]). *Capital: A Critique of Political Economy* (Vol. I). London: Penguin Books.

Massé, F., & Lunstrum, E. (2015). Accumulation by Securitization: Commercial Poaching, Neoliberal Conservation, and the Creation of New Wildlife Frontiers. *Geoforum, 69*, 227–237.

Massé, F., Lunstrum, E., & Holterman, D. (2017). Linking Green Militarization and Critical Military Studies. *Critical Military Studies, 4*(2), 201–221.

Matisons, M. R., & Ross, A. R. (2015). Against Deep Green Resistance. Retrieved from https://www.counterpunch.org/2015/08/11/against-deep-green-resistance/.

Mbembé, A. (2003). Necropolitics. *Public Culture, 15*, 11–40.

McGinnis, M. V. (1999). *Bioregionalism*. London: Routledge.

McKay, B. M. (2017). Agrarian Extractivism in Bolivia. *World Development, 97*, 199–211.

McKittrick, K. (2011). On Plantations, Prisons, and a Black Sense of Place. *Social & Cultural Geography, 12*(8), 947–963.

McMichael, P. (2012). The Land Grab and Corporate Food Regime Restructuring. *The Journal of Peasant Studies, 39*(3–4), 681–701.

McMichael, P. (2013). *Food Regimes and Agrarian Questions*. Rugby: Fernwood Publishing and Practical Action Publishing.

McNally, D. (2012). *Monsters of the Market: Zombies, Vampires, and Global Capitalism*. Chicago: Haymarket Books.

McNeish, J.-A., Borchgrevink, A., & Logan, O. (2015). *Contested Powers: The Politics of Energy and Development in Latin America*. London: Zed Books Ltd.

Merchant, C. (1980). *The Death of Nature: Women, Ecology, and the Scientific Revolution*. New York: Harper & Row.

Merchant, C. (1983). *The Death of Nature*. New York: Harper & Row.

Middeldorp, N., Morales, C., & van der Haar, G. (2016). Social Mobilisation and Violence at the Mining Frontier: The Case of Honduras. *The Extractive Industries and Society, 3*, 930–938.

Mies, M., & Shiva, V. (2014 [1993]). *Ecofeminism*. London: Zed Books.

Milne, D. (2007). "Our Equivalent of Guerrilla Warfare": Walt Rostow and the Bombing of North Vietnam, 1961–1968. *The Journal of Military History, 71*, 169–203.

Mintz, S. (1985). *Sweetness and Power*. New York: Viking Penguin.

Montenegro de Wit, M. (2017). Beating the Bounds: How Does 'Open Source' Become a Seed Commons? *The Journal of Peasant Studies, 46*, 1–36.
Moore, B. (1966). *Social Origins of Dictatorship and Democracy: Lord and Peasant in the Making of the Modern World*. London: Beacon Press.
Moore, J. (2004). *I Am Not a Man, I Am Dynamite!: Friedrich Nietzsche and the Anarchist Tradition*. New York: Autonomedia.
Moore, J. W. (2010). 'Amsterdam Is Standing on Norway' Part I. *Journal of Agrarian Change, 10*(1), 33–68.
Moore, J. W. (2015). *Capitalism in the Web of Life: Ecology and the Accumulation of Capital*. London: Verso.
Moses, A. D. (2002). Conceptual Blockages and Definitional Dilemmas in the 'Racial Century': Genocides of Indigenous Peoples and the Holocaust. *Patterns of Prejudice, 36*, 7–36.
Moses, A. D. (2008). *Empire, Colony, Genocide: Conquest, Occupation, and Subaltern Resistance in World History (War and Genocide)*. Oxford: Berghahn.
Moses, D., & Stone, D. (2013). *Colonialism and Genocide*. London: Routledge.
MTC (Mauvaise Troupe Collective). (2018 [2016]). *The Zad and NoTAV: Territorial Struggles and the Making of a New Political Intelligence*. London: Verso Books.
Naess, A. (1973). The Shallow and the Deep, Long-Range Ecology Movement. A Summary. *Inquiry, 16*(1–4), 95–100.
Neimark, B. (2012). Green Grabbing at the 'Pharm' Gate: Rosy Periwinkle Production in Southern Madagascar. *Journal of Peasant Studies, 39*, 423–445.
Newman, S. (2000). Anarchism and the Politics of Ressentiment. *Theory and Event, 4*, 3–34. Retrieved from https://theanarchistlibrary.org/library/saul-newman-anarchism-and-the-politics-of-ressentiment.
Newman, S. (2001). War on the State: Stirner and Deleuze's Anarchism. *Anarchist Studies, 9*, 147–164.
Newman, S. (2003). Stirner and Foucault: Toward a Post-Kantian Freedom. *Postmodern Culture, 13*(2). https://doi.org/10.1353/pmc.2003.0007.
Newman, S. (2011). *Max Stirner*. New York: Palgrave.
Nibert, D. A. (2013). *Animal Oppression and Animal Violence: Domesecration, Capitalism, and Global Conflict*. New York: Columbia University Press.
Nielsen, K. B. (2018). *Land Dispossession and Everyday Politics in Rural Eastern India*. London and New York: Anthem Press.
Nirmal, P., & Rocheleau, D. (2019). Decolonizing Degrowth in the Post-Development Convergence: Questions, Experiences, and Proposals from Two Indigenous Territories. *Environment and Planning E: Nature and Space*, 1–28.
Nixon, R. (2011). *Slow Violence and the Environmentalism of the Poor*. Cambridge: Harvard University Press.
Noske, B. (1989). *Humans and Other Animals: Beyond the Boundaries of Anthropology*. London: Pluto Press.

O'Brochta, W. (2019). A Meta-Analysis of Natural Resources and Conflict. *Research & Politics, 6*(1). https://doi.org/10.1177/2053168018818232.

O'Connor, J. (1988). Capitalism, Nature, Socialism: A Theoretical Introduction. *Capitalism Nature Socialism, 1*, 1–38.

OECD. (2017). *Health at a Glance: 2017 OECD Indicators*. Retrieved from https://www.oecd-ilibrary.org/social-issues-migration-health/health-at-a-glance-2017/antidepressant-drugs-consumption-2000-and-2015-or-nearest-year_health_glance-2017-graph181-en.

Ojeda, D. (2012). Green Pretexts: Ecotourism, Neoliberal Conservation and Land Grabbing in Tayrona National Natural Park, Colombia. *The Journal of Peasant Studies, 39*, 357–375.

Oliveira, G., & Hecht, S. (2016). Sacred Groves, Sacrifice Zones and Soy Production: Globalization, Intensification and Neo-Nature in South America. *The Journal of Peasant Studies, 43*(2), 251–285.

Paley, D. (2014). *Drug War Capitalism*. Oakland, CA: AK Press.

Peet, R., & Watts, M. (Eds.). (1996). *Liberation Ecologies: Environment, Development, Social Movements*. London and New York: Routledge.

Peet, R., Robbins, P., & Watts, M. (2011). *Global Political Ecology*. New York: Routledge.

Pellow, D. N. (2014). *Total Liberation: The Power and Promise of Animal Rights and the Radical Earth Movement*. Minneapolis, MN and London: University of Minnesota Press.

Peluso, N. L. (1993). Coercing Conservation? The Politics of State Resource Control. *Global Environmental Change, 3*, 199–217.

Peluso, N. L. (2017). Plantations and Mines: Resource Frontiers and the Politics of the Smallholder Slot. *The Journal of Peasant Studies, 44*, 834–869.

Peluso, N. L., & Lund, C. (2011). New Frontiers of Land Control: Introduction. *The Journal of Peasant Studies, 38*, 667–681.

Peluso, N. L., & Vandergeest, P. (2011). Political Ecologies of War and Forests: Counterinsurgencies and the Making of National Natures. *Annals of the Association of American Geographers, 101*, 587–608.

Peluso, N. L., & Watts, M. (2001). *Violent Environments*. Ithaca, NY: Cornell University Press.

Pendleton, M. R. (1998). Policing the Park: Understanding Soft Enforcement. *Journal of Leisure Research, 30*(4), 552–571. https://doi.org/10.1080/00222216.1998.11949847.

Perelman, M. (2007). Primitive Accumulation from Feudalism to Neoliberalism. *Capitalism Nature Socialism, 18*, 44–61.

Perlman, F. (2010 [1983]). *Against His-Story, Against Leviathan*. Detroit: Red and Black Press.

Perreault, T. (2018). The Plantation and the Mine: Comment on "After the Land Grab: Infrastructural Violence and the 'Mafia System' in Indonesia's Oil Palm Plantation Zone" by Tania Li. *Geoforum, 96*, 345–347.

Perreault, T., Bridge, G., & McCarthy, J. (2015). *The Routledge Handbook of Political Ecology*. London: Routledge.

Petras, J., & Veltmeyer, H. (2014a). *Extractive Imperialism in the Americas*. London: Brill.

Petras, J., & Veltmeyer, H. (2014b). *Extractive Imperialism in the Americas: Capitalism's New Frontier*. London: Brill.

Phillips, M. (1993). Rural Gentrification and the Processes of Class Colonisation. *Journal of Rural Studies, 9*, 123–140.

Platt, J. (2013). *Why Winona LaDuke Is fighting for Food Sovereignty*. Retrieved from http://www.mnn.com/leaderboard/stories/why-winona-laduke-isfighting-for-food-sovereignty.

Polanyi, K. (2001 [1944]). *The Great Transformation: The Political and Economic Origins of Our Time*. Boston: Beacon Press.

Porter, G., & Kakabadse, N. K. (2006). HRM Perspectives on Addiction to Technology and Work. *Journal of Management Development, 25*, 535–560.

Post, H. (2015). *The Dark Side of the Sun: Solar Power and Global Electronic Waste*. Retrieved from https://www.huffingtonpost.com/peter-schwartz/solar-power-and-the-looming_b_7104058.html.

Pratt, L. A., Brody, D. J., & Gu, Q. (2017). *Antidepressant Use Among Persons Aged 12 and Over: United States, 2011–2014*. NCHS Data Brief. Number 283. National Center for Health Statistics Brief 283. Retrieved from https://www.cdc.gov/nchs/data/databriefs/db283.pdf.

Preston, J. (2017). Racial Extractivism and White Settler Colonialism: An Examination of the Canadian Tar Sands Mega-Projects. *Cultural Studies, 31*, 353–375.

Price, D. H. (2011). *Weaponizing Anthropology: Social Science in Service of the Militarized State*. Oakland, CA: AK Press and Counterpunch Books.

Price, D. H. (2014). Counterinsurgency by Other Names: Complicating Humanitarian Applied Anthropology in Current, Former, and Future War Zones. *Human Organization, 73*, 95–105.

Rahnema, M. (1997). Development and the People's Immune System: The Story of Another Variety of AIDS. In M. Rahnema & V. Bawtree (Eds.), *The Post-Development Reader* (pp. 111–129). London: Zed Books.

Rahnema, M., & Bawtree, V. (1997). *The Post-Development Reader*. London: Zed Books.

Rajak, D. (2011). *In Good Company: An Anatomy of Corporate Social Responsibility*. Palo Alto, CA: Stanford University Press.

Rasmussen, M. B., & Lund, C. (2018). Reconfiguring Frontier Spaces: The Territorialization of Resource Control. *World Development, 101*, 388–399.

Reclus, E. (2013 [1905]). *Anarchy, Geography, Modernity: Selected Writings of Elisée Reclus*. Oakland, CA: PM Press.

RF. (2016). *Colonisation*. Retrieved from https://en-contrainfo.espiv.net/files/2016/06/Colonisationreturn-fire-vol3.pdf.

RF (Return Fire). (2013–Present). *Return Fire: Anti-Authority-Daily Revolt Individual Will-De-Civilisation*. Retrieved from https://325.nostate.net/distro/.

Ribot, J., & Peluso, N. (2003). A Theory of Access. *Rural Sociology, 68*, 153–181.

Rich, P. B., & Duyvesteyn, I. (2012). The Study of Insurgency and Counterinsurgency. In *The Routledge Handbook of Insurgency and Counterinsurgency* (pp. 1–19). London: Routledge.

Rignall, K. E. (2016). Solar Power, State Power, and the Politics of Energy Transition in Pre-Saharan Morocco. *Environment and Planning A, 48*(3), 540–557.

Robbins, P. (2012 [2004]). *Political Ecology: A Critical Introduction*. Malden, MA: John Wiley & Sons.

Rodgers, D., & O'Neill, B. (2012). Infrastructural Violence: Introduction to the Special Issue. *Ethnography, 13*, 401–412.

Rodney, W. (2009 [1972]). *How Europe Underdeveloped Africa*. Washington, DC: Howard University Press.

Romanyshyn, R. (1989). *Technology as Symptom and Dream*. London: Routledge.

Rosales, A. (2016). Deepening Extractivism and Rentierism: China's Role in Venezuela's Bolivarian Developmental Model. *Canadian Journal of Development Studies/Revue canadienne d'études du développement, 37*(4), 560–577.

Rosset, P. M., & Altieri, M. A. (2017). *Agroecology: Science and Politics*. Rugby: Practical Action Publishing.

Rosset, P., & Martínez-Torres, M. E. (2012). Rural Social Movements and Agroecology: Context, Theory, and Process. *Ecology and Society, 17*, 17–29.

Rosset, P. M., Machín Sosa, B., Roque Jaime, A. M., et al. (2011). The Campesino-to-Campesino Agroecology Movement of ANAP in Cuba: Social Process Methodology in the Construction of Sustainable Peasant Agriculture and Food Sovereignty. *The Journal of Peasant Studies, 38*, 161–191.

Rostow, W. (1960). *The Stages of Economic Growth: A Non-Communist Manifesto*. Cambridge: Cambridge University Press.

Rousselle, D., & Evren, S. (2011). *Post-Anarchism: A Reader*. London: Pluto Press.

Sale, K. (1991 [1985]). *Dwellers in the Land: The Bioregional Vision*. Athens: University of Georgia Press.

Salemink, O. (2003). Social Science Intervention: Moral Versus Political Economy and the Vietnam War. In P. Q. Van Ufford & A. K. Giri (Eds.), *A Moral Critique of Development: In Search of Global Responsibilities*. London: Routledge.

Sarant, L. (2015). *Desertec: An Aborted Project or Just a Change of Direction?* Retrieved from https://www.natureasia.com/en/nmiddleeast/article/10.1038/nmiddleeast.2015.4.

Sayer, D. (1987). *The Violence of Abstraction: The Analytical Foundations of Historical Materialism.* New York: Basil Blackwell.

Scheidel, A., Temper, L., Demaria, F., et al. (2017). Ecological Distribution Conflicts as Forces for Sustainability: An Overview and Conceptual Framework. *Sustainability Science, 13*, 585–598.

Scheper-Hughes, N. (1992). *Death Without Weeping. The Violence of Everyday Life in Brazil.* Berkeley, Los Angeles, and Oxford: University of California Press.

Schoenberger, L., Hall, D., & Vandergeest, P. (2017). What Happened When the Land Grab Came to Southeast Asia? *Journal of Peasant Studies, 44*, 697–725.

Scoones, I., Hall, R., Borras, S. M., et al. (2013). The Politics of Evidence: Methodologies for Understanding the Global Land Rush. *Journal of Peasant Studies, 40*, 469–483.

Scott, J. C. (1977). *The Moral Economy of the Peasant: Rebellion and Subsistence in Southeast Asia.* New Haven: Yale University Press.

Scott, J. C. (1998). *Seeing Like a State: How Certain Schemes to Improve the Human Condition Have Failed.* New Haven, CT: Yale University Press.

Scott, J. C. (2009). *The Art of Not Being Governed: An Anarchist History of Upland Southeast Asia.* New Haven, CT: Yale University Press.

Scott, J. C. (2012). *Two Cheers for Anarchism: Six Easy Pieces on Autonomy, Dignity, and Meaningful Work and Play.* Princeton, NJ: Princeton University Press.

Scott, J. C. (2017). *Against the Grain: A Deep History of the Earliest States.* New Haven, CT: Yale University Press.

Seagle, C. (2012). Inverting the Impacts: Mining, Conservation and Sustainability Claims Near the Rio Tinto/Qmm Ilmenite Mine in Southeast Madagascar. *Journal of Peasant Studies, 39*, 447–477.

Seetharaman, G., & Sharma, S. N. (2018). Find Out How the Government's Marquee Infrastructure Projects Are Doing. *The Economic Times.* Retrieved from https://economictimes.indiatimes.com/news/economy/infrastructure/find-out-how-the-governments-marquee-infrastructure-projects-are-doing/articleshow/65273342.cms?from=mdr.

Shah, A. (2013). The Agrarian Question in a Maoist Guerrilla Zone. *Journal of Agrarian Change, 13*, 424–450.

Sharma, A., & Gupta, A. (2006). *The Anthropology of the State: A Reader.* London: John Wiley & Sons.

Shiva, V. (2002 [1989]). *Staying Alive: Women, Ecology and Development.* London: Zed Books.

Short, D. (2010). Cultural Genocide and Indigenous Peoples: A Sociological Approach. *The International Journal of Human Rights, 14*, 833–848.

Short, D. (2016). *Redefining Genocide: Settler Colonialism, Social Death and Ecocide.* London: Zed Books.

Siamanta, Z. C. (2017). Building a Green Economy of Low Carbon: The Greek Post-Crisis Experience of Photovoltaics and Financial 'Green Grabbing'. *Journal of Political Ecology, 24*, 258–276.

Siamanta, Z. C. (2019). Wind Parks in Post-Crisis Greece: Neoliberalisation Visà-Vis Green Grabbing. *Environment and Planning E: Nature and Space*. https://doi.org/10.1177/2514848619835156.

Siamanta, C., & Dunlap, A. (forthcoming). *'Accumulation by Wind Energy': Wind Energy Development as a Capitalist Trojan Horse in Crete*. Greece and Oaxaca, Mexico: ACME.

Smil, V. (2016). *To Get Wind Power You Need Oil*. Retrieved from http://spectrum.ieee.org/energy/renewables/to-get-wind-power-you-need-oil.

Smith, N. (2010 [1983]). *Uneven Development: Nature, Capital, and the Production of Space*. London: Verso.

Smith, D. P. (2011). What Is Rural Gentrification? Exclusionary Migration, Population Change, and Revalorised Housing Market. *Interface: Planning Theory & Practice, 12*, 596–605.

Smith, P. (2014). *Soaring Copper Prices Drive Wind Farm Crime*. Retrieved from http://www.windpowermonthly.com/article/1281864/soaring-copper-prices-drive-wind-farm-crime.

Solovey, M. (2001). Project Camelot and the 1960s Epistemological Revolution: Rethinking the Politics-Patronage-Social Science Nexus. *Social Studies of Science, 31*, 171–206.

Spivak, G. (1988). Can the Subaltern Speak? In C. Nelson & L. Grossberg (Eds.), *Marxism and the Interpretation of Culture* (pp. 271–313). Urbana: University of Illinois Press.

Springer, S. (2016a). *The Anarchist Roots of Geography: Toward Spatial Emancipation*. Minneapolis, MN: University of Minnesota Press.

Springer, S. (2016b). *Learning Through the Soles of Our Feet: Unschooling, Anarchism, and the Geography of Childhood*. London: Rowman & Littlefield.

Springer, S. (2017). The Limits to Marx: David Harvey and the Condition of Postfraternity. *Dialogues in Human Geography, 7*(3), 280–294.

Springer, S. (2019). Total Liberation Ecology: Integral Anarchism, Anthroparchy, and the Violence of Indifference. In S. Springer, M. Locret, J. Mateer, et al. (Eds.), *Anarchist Political Ecology*. Oakland, CA: PM Press.

Springer, S., & Le Billon, P. (2016). Violence and Space: An Introduction to the Geographies of Violence. *Political Geography, 52*, 1–3.

Springer, S., Locret, M., Mateer, J., et al. (2019). *Anarchist Political Ecology*. Oakland, CA: PM Press.

Stephen, L. (2000). The Construction of Indigenous Suspects: Militarization and the Gendered and Ethnic Dynamics of Human Rights Abuses in Southern Mexico. *American Ethnologist, 26*, 822–842.

Stirner, M. (2017 [1845]). *The Unique and Its Property*. Berkeley, CA: Ardent Press.

Strong, Z. (2016). *Will the Transition to Renewable Energy Be Paved in Copper?* Retrieved from https://www.renewableenergyworld.com/articles/2016/01/will-the-transition-to-renewable-energy-be-paved-in-copper.html.

Stuart, D., Schewe, R. L., & Gunderson, R. (2013). Alienation in the Dairy Sector. *Sociologia Ruralis, 53*, 201–222.

Sullivan, S. (2006). Elephant in the Room? Problematising 'New' (Neoliberal) Biodiversity Conservation. *Forum for Development Studies, 33*, 105–135.

Sullivan, S. (2009). Green Capitalism, and the Cultural Poverty of Constructing Nature as Service Provider. *Radical Anthropology, 3*, 18–27.

Sullivan, S. (2010). 'Ecosystem Service Commodities'—A New Imperial Ecology? Implications for Animist Immanent Ecologies, with Deleuze and Guattari. *New Formations: A Journal of Culture/Theory/Politics, 69*, 111–128.

Sullivan, S. (2013a). Banking Nature? The Spectacular Financialisation of Environmental Conservation. *Antipode, 45*, 198–217.

Sullivan, S. (2013b). After the Green Rush? Biodiversity Off-sets, Uranium Power and the 'Calculus of Casualties' in Greening Growth. *Human Geography, 6*, 80–101.

Sullivan, S. (2017a). What's Ontology Got to Do with It? On Nature and Knowledge in a Political Ecology of the 'Green Economy'. *Journal of Political Ecology, 24*, 217–242.

Sullivan, S. (2017b). On 'Natural Capital', 'Fairy Tales' and Ideology. *Development and Change, 48*, 397–423.

Sundar, N. (2016). *The Burning Forest: India's War in Bastar*. New Delhi: Juggernaut.

Svampa, M. (2013 [2011]). Resource Extractivism and Alternatives: Latin American Perspectives on Development. In M. Lang & D. Mokrani (Eds.), *Beyond Development: Alternative Visions from Latin America* (pp. 117–144). Amsterdam: Transnational Institute.

Svarstad, H., Benjaminsen, T. A., & Overå, R. (2018). Power Theories in Political Ecology. *Journal of Political Ecology, 25*, 350–363.

Swyngedouw, E. (1996). The City as a Hybrid: On Nature, Society and Cyborg Urbanization. *Capitalism Nature Socialism, 7*(2), 65–80.

Tabassum-Abbasi, P. M., Abbasi, T., et al. (2014). Wind Energy: Increasing Deployment, Rising Environmental Concerns. *Renewable and Sustainable Energy Reviews, 31*, 270–288.

Taussig, M. T. (1980). *The Devil and Commodity Fetishism in South America*. Durham, NC: University of North Carolina Press.

Tetreault, D. V. (2014). Mexico: The Political Ecology of Mining. In H. Veltmeyer & J. F. Petras (Eds.), *The New Extractivism: A Post-Neoliberal Development Model or Imperialism of the Twenty-First Century?* London: Zed Books.

Thayer, R. L. (2003). *LifePlace: Bioregional Thought and Practice*. Berkeley, CA: University of California Press.

Thomas, W. (1994). *Scorched Earth: Military's Assault on the Environment.* Philadelphia: New Society Publishers.

Thompson, E. P. (1990 [1975]). *Whigs and Hunters.* London: Penguin Books.

Thompson, E. P. (1991 [1963]). *The Making of the English Working Class.* London: Penguin Books.

Thone, F. (1935). Nature Rambling: We Fight for Grass. *The Science Newsletter, 27*, 14.

Thoreau, H. D. (1971 [1854]). *Walden.* Princeton, NJ: Princeton University Press.

TIC (The Invisible Committee). (2015). *To Our Friends.* South Pasadena: Semiotext(e).

TIQQUN. (2011 [2001]). *The Cybernetic Hypothesis.* TIQQUN 2. Retrieved from https://theanarchistlibrary.org/library/tiqqun-the-cybernetic-hypothesis.

Torres, B. (2005). *Making a Killing: The Political Economy of Animal Rights.* Oakland, CA: AK Press.

Torres, B. (2007). *Making a Killing: The Political Economy of Animal Rights.* Oakland: AK Press.

Twine, R. (2012). Revealing the 'Animal-Industrial Complex'—A Concept & Method for Critical Animal Studies? *Journal for Critical Animal Studies, 10*(1), 12–39.

UCS. (2013). *Environmental Impacts of Solar Power.* Retrieved from https://www.ucsusa.org/clean_energy/our-energy-choices/renew-able-energy/environmental-impacts-solar-power.html#.W-NS9TFMHVk.

Ulloa, A. (2013 [2005]). *The Ecological Native: Indigenous Peoples' Movements and Eco-Governmentality in Columbia.* London: Routledge.

UNSDG. (2018). *Sustainably Manage Forests, Combat Desertification, Halt and Reverse Land Degradation, Halt Biodiversity Loss.* Retrieved from https://www.un.org/sustainabledevelopment/biodiversity/.

UNSDG. (2019). UN Report: Nature's Dangerous Decline 'Unprecedented'; Species Extinction Rates 'Accelerating'. Retrieved from https://www.un.org/sustainabledevelopment/blog/2019/05/nature-decline-unprecedented-report/.

Uribe, S. (2018). Illegible Infrastructures: Road Building and the Making of State-Spaces in the Colombian Amazon. *Environment and Planning D: Society and Space*, 1–19.

Vela-Almeida, D. (2018). Territorial Partitions, the Production of Mining Territory and the Building of a Post-Neoliberal and Plurinational State in Ecuador. *Political Geography, 62*, 126–136.

Veltmeyer, H. (2013). The Political Economy of Natural Resource Extraction: A New Model or Extractive Imperialism? *Canadian Journal of Development Studies/Revue canadienne d'études du développement, 34*, 79–95.

Veltmeyer, H., & Petras, J. F. (2014). *The New Extractivism: A Post-Neoliberal Development Model or Imperialism of the Twenty-First Century?* London: Zed Books.

Veracini, L. (2014). Understanding Colonialism and Settler Colonialism as Distinct Formations. *Interventions, 16*, 615–633.

Vergara-Camus, L. (2009). The MST and the EZLN Struggle for Land: New Forms of Peasant Rebellions. *Journal of Agrarian Change, 9*, 365–391.

Vergara-Camus, L. (2014). *Land and Freedom: The MST, the Zapatistas and Peasant Alternatives to Neoliberalism*. London: Zed Books.

Vergara-Camus, L., & Kay, C. (2017). Agribusiness, Peasants, Left-Wing Governments, and the State in Latin America: An Overview and Theoretical Reflections. *Journal of Agrarian Change, 17*, 239–257.

Vergara-Camus, L., & Kay, C. (2018). New Agrarian Democracies: The Pink Tide's Lost Opportunity. *Socialist Register, 54*, 224–243.

Verweijen, J., & Dunlap, A. (forthcoming). The Evolving Techniques of Social Engineering, Land Control and Managing Protest Against Extractivism: Introducing Political (Re)Actions 'From Above'. *Political Geography*.

Verweijen, J., & Marijnen, E. (2018). The Counterinsurgency/Conservation Nexus: Guerrilla Livelihoods and the Dynamics of Conflict and Violence in the Virunga National Park, Democratic Republic of the Congo. *The Journal of Peasant Studies, 45*, 300–320.

Vetlesen, A. J., & Willig, R. (2018). *Hva skal vi svare våre barn?* Oslo: Dreyer forlag.

Vidal, J. (2008). *The Great Green Land Grab*. Retrieved from http://www.theguardian.com/environment/2008/feb/13/conservation.

Vijayabaskar, M., & Menon, A. (2018). Dispossession by Neglect: Agricultural Land Sales in Southern India. *Journal of Agrarian Change, 18*, 571–587.

Virilio, P. (2008 [1983]). *Pure War*. Los Angeles: Semiotext(e).

Vorbrugg, A. (2019). Not About Land, Not Quite a Grab: Dispersed Dispossession in Rural Russia. *Antipode*. https://doi.org/10.1111/anti.12523.

Wadiwel, D. (2015). *The War Against Animals*. Leiden: Brill.

Walker, R. A. (2004). *The Conquest of Bread: 150 Years of Agribusiness in California*. New York: The New Press.

Watt, A. (1981). Illich and Anarchism. *Educational Philosophy and Theory, 13*, 1–15.

Watts, M. J. (1983a). On the Poverty of Theory: Natural Hazards Research in Context. In K. Hewitt (Ed.), *Interpretations of Calamity: From the Viewpoint of Human Ecology* (pp. 231–262). Boston: Allen & Unwin.

Watts, M. J. (1983b). *Silent Violence: Food, Famine, and Peasantry in Northern Nigeria*. Berkeley, CA: University of California Press.

Watts, M. (2004). Resource Curse? Governmentality, Oil and Power in the Niger Delta, Nigeria. *Geopolitics, 9*(1), 50–80.

Watts, M., Robbins, P., & Peet, R. (2011). Global Nature. In R. Peet, P. Robbins, & M. Watts (Eds.), *Global Political Ecology*. London and New York: Routledge.

Weber, M. (1992 [1930]). *The Protestant Ethic and the Spirit of Capitalism*. London and New York: Routledge.

Weis, T. (2010). The Accelerating Biophysical Contradictions of Industrial Capitalist Agriculture. *Journal of Agrarian Change, 10*, 315–341.
Weis, T. (2013). *The Ecological Hoofprint: The Global Burden of Industrial Livestock*. London: Zed Books.
Weis, T. (2016). Toward 120 Billion: Dietary Change and Animal Lives. *Radical Philosophy, 199*, 8–13.
Weis, T. (2018). Ghosts and Things: Agriculture and Animal Life. *Global Environmental Politics, 18*(2), 134–142.
Weizman, E. (2011). *The Least of All Possible Evils: Humanitarian Violence from Arendt to Gaza*. London: Verso.
West, P., Igoe, J., & Brockington, D. (2006). Parks and Peoples: The Social Impact of Protected Areas. *Annual Review of Anthropology, 35*, 251–277.
White, R. J. (2017a). Capitalsim and the Commodification of Animals: The Need for Critical Vegan Praxis. In D. Nibert (Ed.), *Animal Oppression and Capitalism*. Connecticut: Praeger.
White, R. J. (2017b). Capitalism and the Commodification of Animals: The Need for Critical Vegan Praxis, Animated by Anarchism! In D. Nibert (Ed.), *Animal Oppression and Capitalism*. Connecticut: Praeger.
White, B., Borras, S. M., Hall, R., et al. (2012). The New Enclosures: Critical Perspectives on Corporate Land Deals. *Journal of Peasant Studies, 39*, 619–647.
Williams, R. (1973). *The Country and the City*. New York: Oxford University Press.
Williams, K. (2007 [2004]). *Our Enemies in Blue: Police and Power in America*. Cambridge: South End Press.
Williams, K., Munger, W., & Messersmith-Glavin, L. (2013). *Life During Wartime: Resisting Counterinsurgency*. Edinburgh: AK Press.
Wilson, E., & Stammler, F. (2016). Beyond Extractivism and Alternative Cosmologies: Arctic Communities and Extractive Industries in Uncertain Times. *The Extractive Industries and Society, 3*, 1–8.
Wolf, E. (1972). Ownership and Political Ecology. *Anthropological Quarterly, 45*, 201–205.
Wolf, E. R. (1999 [1969]). *Peasant Wars of the Twentieth Century*. Omaha: University of Oklahoma Press.
Wolfe, P. (2006). Settler Colonialism and the Elimination of the Native. *Journal of Genocide Research, 8*, 387–409.
Wolford, W., Borras, S. M., Hall, R., et al. (2013). Governing Global Land Deals: The Role of the State in the Rush for Land. *Development and Change, 44*, 189–210.
Wuebbles, D., Fahey, D., Hibbard, K., et al. (2017). *US Global Change Research Program: Climate Science Special Report*. Retrieved from https://science2017.globalchange.gov/chapter/executive-summary/.

Xu, Y. (2019). Politics of Inclusion and Exclusion in the Chinese Industrial Tree Plantation Sector: The Global Resource Rush Seen from Inside China. *The Journal of Peasant Studies, 46*(4), 767–791.

Yang, H.-J., Lim, S.-Y., & Yoo, S.-H. (2017). The Environmental Costs of Photovoltaic Power Plants in South Korea: A Choice Experiment Study. *Sustainability, 9*, 1773.

Ybarra, M. (2012). Taming the Jungle, Saving the Maya Forest: Sedimented Counterinsurgency Practices in Contemporary Guatemalan Conservation. *The Journal of Peasant Studies, 39*, 479–502.

Ybarra, M. (2017). *Green Wars: Conservation and Decolonization in the Maya Forest*. Oakland, CA: University of California Press.

Ye, J., van der Ploeg, J. D., Schneider, S., et al. (2019). The Incursions of Extractivism: Moving from Dispersed Places to Global Capitalism. *The Journal of Peasant Studies*, 1–29. https://doi.org/10.1080/03066150.2018.1559834.

Yenneti, K., Day, R., & Golubchikov, O. (2016). Spatial Justice and the Land Politics of Renewables: Dispossessing Vulnerable Communities Through Solar Energy Mega-Projects. *Geoforum, 76*, 90–99.

Zapatistas. (2016). *Critical Thought in the Face of the Capitalist Hydra I*. Durham: Paper Boat Press.

Zerzan, J. (1988). *Elements of Refusal*. Seattle, WA: C.A.L. PressIPaleo Editions.

Zerzan, J. (2005 [1999]). *Against Civilization: Readings and Reflections*. Port Townsend, WA: Feral House.

Zoomers, A. (2010). Globalisation and the Foreignisation of Space: Seven Processes Driving the Current Global Land Grab. *Journal of Peasant Studies, 37*, 429–447.

Index[1]

A

Academia, 17, 45n1, 61
Agency, 8, 17, 84, 98–100, 125
Agrarian, 5, 8, 9, 36, 44–61, 76, 78, 95, 97, 102, 121
Agriculture, 7, 9, 16, 47, 58, 76, 93–95, 103, 108
Anarchism/anarchist, 20, 32, 47, 54–57, 103, 104, 123, 128
Animals, 7, 10, 15, 32, 35, 101–104, 101n3, 109, 111, 122, 124, 125, 127–129
Anthropocene, 8, 17, 35
Anthropology, 9, 48, 77
Apparatus, 25, 48, 55, 98, 101
Autonomy, 29, 125–126

B

Bookchin, Murray, 15, 55, 104
Bureaucracy, 7, 78

C

Capitalism, 2–10, 14–36, 47, 60, 76, 92, 96, 100, 102, 103, 110, 123, 127
Capitalocene, 94, 100n2
Climate change/catastrophe, 3, 8, 17, 34, 52, 81, 94, 104–106, 129
Colonization, 14, 15, 22, 26, 33, 78, 110
Conflict, 46, 49, 57, 57n6, 58, 75, 76, 128
Consumerism, 25, 33, 34, 80, 96
Counterinsurgency, 9, 48, 60, 76, 78–84, 108, 121
Critical agrarian studies, 5, 8, 9, 36, 44–61, 76, 78, 97, 121
Cybernetic, 7, 25, 31, 35, 124

D

Decolonization, 50, 80
Defenders, land, 100

[1] Note: Page numbers followed by 'n' refer to notes.

A. Dunlap, J. Jakobsen, *The Violent Technologies of Extraction*,
https://doi.org/10.1007/978-3-030-26852-7

Degrowth, 56, 110, 125–128
Desertification, 3
Development, 3, 5, 6, 8, 9, 20, 21, 26, 27, 30–34, 44, 45, 52–53, 56–58, 60, 75, 78, 80–84, 93, 99–101, 100n2, 106, 107, 109, 122–124
Disaster, 6, 15, 16, 55, 127, 128
Divisions of labor, 16, 21, 26, 35, 122, 125
Domination, 4, 16, 18, 19, 21, 32, 35, 55, 76, 77, 103, 104, 111, 123, 128
Dunlap, Alexander, 5, 6, 8, 14, 16, 24, 26, 27, 30, 33, 34, 48–50, 52–55, 58, 59, 75–78, 81–83, 93, 97, 100, 105–110, 123

E

Economic growth, 75, 81, 110, 122
Ecosystem, 2, 3, 27, 58, 93, 123, 128, 130
Ellul, Jacques, 8, 30, 31, 122
Energy (solar, renewable, hydrological), 3, 5, 7, 10, 17, 20, 27, 28, 51, 58, 59, 78, 82, 83, 93, 105–111, 121
Environment, 15, 17, 34, 46, 52, 76, 77, 92, 103, 106, 123, 128, 129

F

Factory farm, 3, 32, 96, 101–105, 111, 121
Finance, 51
Foucault, Michel, 9, 17, 33, 57, 57n5, 75, 78
Franquesa, Jaume, 5, 6

G

Gelderloos, Peter, 8, 21, 29–31, 35, 77
Genocide Machine, 32–35
Geography, 9, 25n8, 51, 77, 126
Global economy, 126
Green economy, 34, 51, 53, 59, 61, 78, 81, 84, 94, 106, 111
Grid, 10, 27–29, 36, 54, 85, 93, 96, 99, 108, 111
Gudynas, Eduardo, 5, 46, 53, 56, 93, 95

H

Habitat, 10, 98, 108, 122, 124, 126, 129, 130
Hierarchy, 29, 32, 55
Hobbes, Thomas, 22, 23
Hydrocarbons, 7, 74

I

Industrial humanity/humans, 2–4, 7, 8, 10, 14–16, 19, 23, 30–32, 34, 50, 61, 84, 96, 98, 101, 102, 104, 121, 122, 124, 125, 127–129
 industrial corridor, 27
Infrastructure, 6, 10, 21, 25, 27, 29, 32, 53, 58, 76, 77, 82, 83, 99, 101, 106, 107, 109, 111, 123, 129
itical agrarian studies, 8

J

Jakobsen, Jostein, 49, 52, 54, 94, 102, 103

K

Knowledge, 2, 7, 7n4, 17, 17n4, 18n6, 26, 48, 49, 55, 60, 77, 82, 121n1, 123, 125n4, 126

L

Land control, 9, 44, 46, 51, 76–79, 81, 83, 84, 95, 100, 105, 107
Land grabbing, 51, 52, 54, 57n6, 58, 78, 84, 103, 108

Leviathan, 19–25, 30
Logistics, 29

M
Manufacture, 98
Marx, Karl, 10, 18, 25, 27, 47, 57, 75
Media, 7n4, 34n12, 81, 127
Military/militarization, 9, 29, 36, 48, 61, 74–85, 121
Mines
  mining, 29, 33, 82, 92, 94, 97, 98, 101, 105–107, 109, 110, 124
  space mining, 101
Modernity, 5, 120, 122, 129

N
Natural resource extraction
  conventional extractivism, 34
  extractive peasant, 98
  extractivism, 5, 95, 121
  green extractivism, 94
  Neo-extractivism, 5
  total extractivism, 5
Neocolonialism, 91
Non-human nature, 14, 23, 58, 75, 85, 123, 130

P
Paramilitaries, 77, 78, 100
Patriarchy, 21, 26, 35, 122
Perlman, Fredy, 7, 8, 10, 19–25, 30, 31, 35, 36, 49, 84, 95, 96, 100
Plantations, 2, 3, 9, 9n6, 10, 16, 19, 20, 29, 45, 58, 76, 92, 96–101, 103, 106, 111
Police, 9, 74, 77, 81, 82, 121, 124, 127
Political Ecology, 5, 8, 9, 36, 44–61, 76, 78–84, 100, 101, 103, 104, 128
Political economy, 4, 33, 45, 53, 79
Post-development, 97, 120
Power, 2, 3, 9, 20–22, 25, 27, 29–31, 34, 35, 46, 55, 57, 76, 78–80, 107, 109–111, 120–130
Profit, 6, 15, 16, 103, 104, 110, 126, 127, 130
Progress, 2–4, 6, 7, 9, 16, 25, 30–34, 36, 44, 49, 56, 60, 61, 74, 93–111, 122, 123, 125
Psychological operations (psy ops), 81

R
Racism, 33, 48, 77, 122, 128
Renewable energy, 10, 59, 93, 96, 106–110
Renewable energy-extraction nexus, 105–111, 121
Repression, 77, 80
Resistance, 25, 33, 34, 46, 56–58, 75, 93, 100, 105

S
Science, 15, 32, 48, 50, 59, 77, 102, 121n1
Scott, James C., 23, 24, 29–31, 34, 48, 54, 74
Self-determination, 54
Short, Damien, 25, 27, 32, 75
Smallholder, 6, 10, 57, 93, 94, 96–98
Social engineering, 49, 83, 84, 97, 104
Solar energy, 106, 107
Spirit, 2, 7, 8, 14–36, 122, 125
Springer, Simon, 16, 20, 55, 56, 58, 59, 77, 103, 104, 125, 128
State, 4–7, 14, 19, 21–24, 26, 27, 29–31, 33, 52, 54, 55, 58, 59, 74–76, 80, 81, 92, 97–101, 103, 122
Stirner, Max, 7, 8, 18, 19, 31, 56, 57, 57n5, 125

T
Taussig, Michael, 3, 7, 16, 18, 18n6, 19, 24, 25, 32, 56, 92, 96, 111, 125
Techno-capitalist progress, 9, 20, 29, 33, 44, 59–61, 84, 121–123
Technology, 4, 6, 7, 7n4, 9, 16, 26, 32, 35, 56, 60, 77, 78, 80n3, 82–84, 83n4, 93, 96–99, 101, 102, 104, 108, 110, 111, 122–125, 127
  technological singularity, 25
Territorialization, 9, 31, 44, 46, 76–79, 81, 83, 84
The Invisible Committee (TIC), 8, 20
Totality, 10, 22, 31, 99, 128
Total liberation, 8, 10, 56, 103, 104, 128, 129

U
United Nations, 3
University, 21, 56, 61

V
Violence
  bureaucratic, 77
  infrastructural, 77, 83, 99
  political, 9, 33, 79, 101
  structural, 77
  symbolic, 77, 78, 83

W
Wind energy, 78, 82, 83, 92, 106, 107, 109, 110
Wolfe, Patrick, 75, 84
Worldeater, 2, 7–10, 14–36, 44–61, 74–76, 78, 79, 83–85, 92–111, 120–125, 125n4, 127–130

Z
Zapatistas (EZLN), 8, 25